Quality Management Essentials

Quality Management Essentials

Ivan Popov
University of Portsmouth, UK

World Scientific

EW JERSEY · LONDON · SINGAPORE · BEIJING · SHANGHAI · HONG KONG · TAIPEI · CHENNAI · TOKYO

Published by

World Scientific Publishing Europe Ltd.

57 Shelton Street, Covent Garden, London WC2H 9HE

Head office: 5 Toh Tuck Link, Singapore 596224

USA office: 27 Warren Street, Suite 401-402, Hackensack, NJ 07601

Library of Congress Cataloging-in-Publication Data

Names: Popov, Ivan (Mechanical engineer), author.

Title: Quality management essentials / Ivan Popov, University of Portsmouth, UK.

Description: New Jersey : World Scientific, [2022] | Includes bibliographical references and index.

Identifiers: LCCN 2021061612 | ISBN 9781800612280 (hardcover) |

ISBN 9781800612297 (ebook for institutions) | ISBN 9781800612303 (ebook for individuals)

Subjects: LCSH: Quality control.

Classification: LCC TS156 .P657 2022 | DDC 658.5/62--dc23/eng20220519

LC record available at https://lccn.loc.gov/2021061612

British Library Cataloguing-in-Publication Data

A catalogue record for this book is available from the British Library.

For any available supplementary material, please visit
https://www.worldscientific.com/worldscibooks/10.1142/Q0362#t=suppl

Desk Editors: Balamurugan Rajendran/Adam Binnie/Shi Ying Koe

Typeset by Stallion Press
Email: enquiries@stallionpress.com

I dedicate this book to my father, Dr Evgeni Popov, who was a noble, strong and smart man.

Preface

He who would learn to fly one day must first learn to walk and run and climb and dance; one cannot fly into flying.

Friedrich Nietzsche

Many good books on quality management are available with general theory or heavy mathematics. This book aims to provide the readers with the most essential practical knowledge and understanding of quality-related issues, so that by making correct decisions faster, they can effectively contribute to a modern dynamic production environment.

Dr Ivan Popov
October 2021

About the Author

Ivan Popov is a Principal Lecturer of Manufacturing Engineering in the School of Mechanical and Design Engineering, University of Portsmouth. He has industrial experience in Engineering Design and Manufacturing. Currently, he is responsible for teaching Quality Management and CADCAM-related modules. His research interests are in Reverse Engineering, Quality Control and Manufacturing Engineering. Dr Popov has been involved in several industrial projects related to Quality Control, Reverse Engineering and CADCAM systems. He has many years of experience in Dimensional Metrology. Dr Popov holds a MEng and PhD in Mechanical Engineering. He is a Chartered Engineer, a Member of the Institution of Engineering and Technology (IET), and a Fellow of the Higher Education Academy. He has published more than 15 journal papers and conference proceedings on Quality Control, Manufacturing Engineering, CADCAM and Reverse Engineering.

Contents

List of Figures

List of Tables

Chapter 1

Introduction to Quality Management

1.1 What is Quality?

Although the concept of quality has existed from early times, the study and definition of quality have been given prominence only in the last century. Following the Industrial Revolution and the development of mass production, it became important to manage the quality of products. Initially, the main goal of quality management was to ensure that engineering requirements were met in the final products. Later, as manufacturing processes became more complex, quality developed into a discipline for controlling process variation as a means of producing quality products.

The use of the term 'quality' might be very subjective. Different people may have different perceptions of it, which often causes confusion or misunderstanding.

1.1.1 Definitions for quality

Quality has been defined as fit for purpose, conformance to requirements, the pursuit of excellence, etc.

- Fit for use:
  This is one of the shortest definitions. It gives a general sense of the meaning but is rather limited in scope.

1

- Meeting customer requirements at a minimum cost:
  This is a very popular definition often associated with the phrase 'good value for money'. Although these two are not quite the same, they both open the question: is the cost a quality attribute?

- Degree to which a set of inherent characteristics of a good or service fulfils customer's stated or implied requirements [1]:
  The last one is considered the most comprehensive definition, covering the widest aspects of the term 'quality'. For example, a customer might be pleasantly surprised by some unexpected features that a product may have (in addition to those initially required) implied by the sales man.

Any organisation focused on quality should promote a culture that results in the behaviour, attitudes, activities and processes that deliver value through fulfilling the needs and expectations of the customer.

The quality of an organisation's products and services is determined by the ability to satisfy customers and the intended and unintended impact on relevant interested parties.

The quality of a product or service includes not only their intended function and performance but also their perceived value and benefit to the customer. For example, a simple calculator with very basic functions and performance might be of great value for someone who has limited needs, while someone else would be very unsatisfied with such a product and would prefer a computer.

1.1.2 *Why is the quality of a product or service so important?*

Nowadays, all producers face certain challenges in their business environment: economic crisis, globalisation as well as severe competition in its market. A company that does not maintain a competitive quality has no chance of staying in the market.

1.1.3 *Quality 'Gurus'*

Many prominent scientists have emerged within the quality field, but some have stood out as key figures of quality. Even though most have passed away, their memories still live on in the ideas, concepts and methods that form our quality thinking today.

In no particular order, the top seven quality gurus who have greatly contributed to quality as we know it today are as follows:

- Walter Shewhart: Shewhart's control charts are widely used to monitor processes. Problems are framed in terms of special cause (assignable cause) and common cause (chance cause). He is referred to as the 'Father of Statistical Quality Control'.
- William Deming, an American engineer, statistician and lecturer: He is famous for his plan-do-study-act (PDSA) cycle, which he developed working on the original Shewhart's plan-do-check-act (PDCA) cycle. Deming is also known for his '14 points for Total Quality Management' originally presented in 'Out of Crisis' [2], which served as management guidelines.
- Joseph Juran, often considered as the 'Father of Quality Management': He developed the principles of internal customer, continuous improvement and cost of quality.
- Philip Crosby: He developed both the 'zero defect' and 'doing it right from the first time' concepts. He suggested the idea of 'The Four Absolutes of Quality Management'.
- Armand Feigenbaum: He originated the concept of total quality management (TQM). He said, 'Quality is everybody's job, but because it is everybody's job, it can become nobody's job without the proper leadership and organisation'.
- Kaoru Ishikawa, best known for fishbone/cause and effect, seven basic tools of quality and quality circles: He is known as the father of Japanese quality control effort. He established the concept of company wide quality control (CWQC) — participation from the top to the bottom of an organisation and from the start to the finish of the product life cycle.
- Genichi Taguchi, a Japanese engineer and statistician: He developed the ideas of robust design and quality loss function. Taguchi recommends a three-stage design process: System design (stage 1): development of a basic functional prototype design, determination of materials, parts and assembly system, determination of the manufacturing process involved. Parameter design (stage 2): selecting the nominals of the system by running statistically planned experiments Design for Six Sigma/Design of Experiment (DFSS/DOE). Tolerance design (stage 3): deals with tightening tolerances and upgrading materials.

Taguchi also contributed to some fundamental innovations in design of experiment.

1.2 What is Quality Management?

According to ISO 9000, quality management (QM) is a set of coordinated activities to direct and control an organisation with respect to quality.

There are three main managerial processes in the QM (Figure 1.1).

- Quality planning: one of the gurus of QM, J. Juran, stated that quality does not happen just like that, it must be planned. According to ISO 9001, there are two main elements of planning that have to be addressed:
  o Quality objectives: generally stated in the quality policy document and incorporated in the product requirements.
  o QM system planning (see Chapter 10).
- Quality control: this is the core of QM and hence it will be covered in greater detail in this book.
- Quality improvement: an important part of QM, which is essential for the sustainability of any business. Without quality improvement, even the best product/service will die sooner or later.

1.3 Quality Control (QC)

1.3.1 *What is QC?*

QC is universal managerial process for conducting operations so as to provide stability to prevent adverse change and to maintain the status quo.

Josef M. Juran

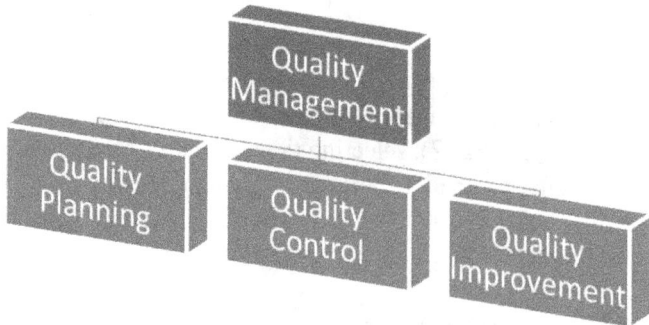

Figure 1.1.   Three main managerial processes in QM.

1.3.2 *How does QC work?*

To achieve quality and to maintain stability, the QC process **evaluates** actual performance, **compares** it to goals and **acts** on the difference.

1.3.3 *QC process*

The flowchart in Figure 1.2 shows the general workflow of the QC process.

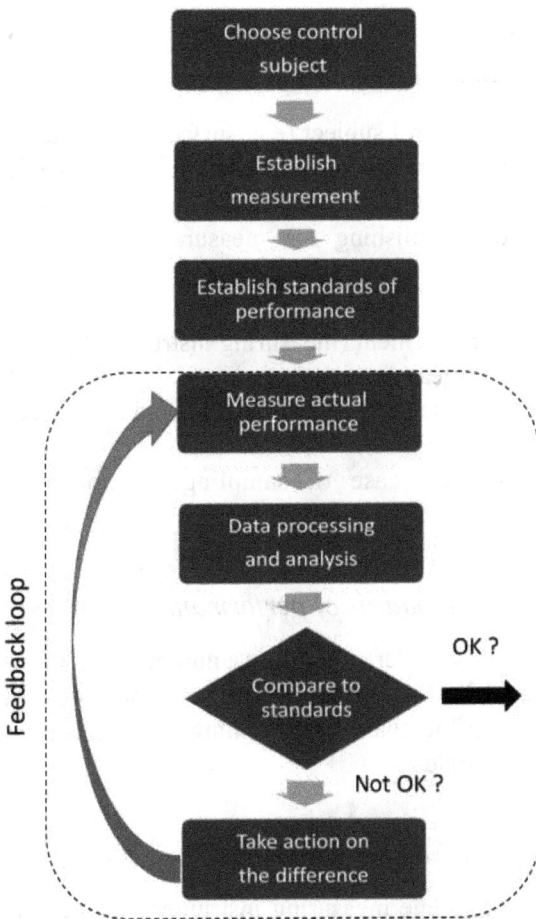

Figure 1.2. QC process.

1.3.3.1 *Choose the control subject*

This is the critical first step. Any essential feature of a product (good or service) may become a control subject—a centre around which the feedback loop is built. Control subjects are derived from multiple sources, which may include:

- stated customer demand for product features;
- technological analysis to translate customer demand into product and process features (see later chapters);
- relevant standards including health, safety and environmental protection.

1.3.3.2 *Establish means of measurement*

After selecting the control subject (e.g. surface finish), the next step is to establish the means of measuring the actual performance of the process or the quality level of the product. Measurement is one of the most difficult tasks in QM. In establishing the measurement, we need to clearly specify:

- the means of measurement (measuring instrument);
- the frequency of measurement;
- the format of the data (in this example, it could be units, type of surface finish parameter, etc.);
- data processing (in case of sampling) to convert it to usable information.

1.3.3.3 *Establish standards of performance*

The standards could be tolerances, limits, numbers (of defects), etc. In this example, it could be the surface finish limit measured in micrometre. Undoubtedly, the prime goal of establishing the standards is to guarantee the quality of the product.

1.3.3.4 *Measure actual performance*

Generally, we need some measuring instrument/device, but we can also count or use some of the human senses: hearing, vision, taste, sense of smell, touch, etc.

1.3.3.5 *Process and analyse data*

Often, the data contain multiple measurements/counts. As data serve as the basis for action, in order to extract the correct information, clarity is key. They must be well organised, processed and analysed through the use of statistical methods. This may include calculating some basic statistics, such as average, standard deviation, checking for outliers (see following chapters) and distribution.

1.3.3.6 *Compare to standards*

The act of comparing to standards is often seen as the role of an umpire. The umpire may be a human being or a technological device. After this comparison, a decision on the action taken should be generated.

1.3.3.7 *Act on the difference*

This may solely require an adjustment of settings, but it could be more than that. The action may require more radical measures, such as a change of tool, machine, part or even the whole technology. In this example, it might involve a change of the finishing process.

1.3.4 *Some important terms used in QC*

Let us clarify the following terms:

1.3.4.1 *Test*

Determination of one or more characteristics according to a procedure.

Inspection: conformity evaluation by observation, measurement, testing or gauging.
Verification: confirmation through the provision of objective evidence that *specified requirements* have been fulfilled.
Validation: confirmation through the provision of objective evidence that the *requirements for a specific intended use or application* have been fulfilled.

Activity 1.1: In the context of QM, identify which of the above terms applies to each of the following cases:

1. Fire a set of ammunition to approve their production process.
2. Conduct an experiment aiming to identify the maximum load/weight a cardboard box can sustain.
3. As a quality manager, suggest a plan for staff training.
4. Find the weight of a brick and compare against a limit.
5. Compare two sets of documents.

1.4 Quality Assurance

The terms 'Quality Assurance' (QA) and 'Quality Control' are often used interchangeably to refer to ways of ensuring the quality of a service or product. QA can be defined [1] as 'part of QM focused on *providing confidence* that quality requirements will be fulfilled'.

These terms have some commonalities and distinctions as well. Both evaluate and compare performances to the goals and act on the difference. However, QA aspects of QM have primary focus on *demonstrating and providing confidence,* ensuring that the requirements for quality will be achieved. While QA relates to how a process is performed or how a product is made, QC is more about the inspection aspect of QM. The confidence provided by QA is bidirectional—internally to management and externally to customers, regulators, certifiers and third parties.

Chapter 2

Measurements for Quality Assurance

Without measurements, any machine would be just a garbage making tool.

T. Toshkov

2.1 What is Measurement?

There are several similar definitions for this, but a simple one states that **measurement** is a comparison between a standard unit (e.g. metre, m) and an unknown quantity of interest—measurand (e.g. 1.5 m).

2.2 Methods of Measurement

2.2.1 *Direct or indirect*

Direct measurement directly identifies the targeted value. An example is measuring the size of a piece of furniture using a measuring tape.

Indirect measurement would directly measure another parameter/unit and then, using a formula/graph (a known relationship) would allow the identification of the targeted value. An example is measuring the power consumption of an electric motor by using the formula $P = I \cdot V$, in W, and directly measuring the electric current I in A and the voltage V in V.

2.2.2 *Absolute or relative*

Absolute is a measurement resulting in the whole magnitude of the unknown value. An example is length of a bolt measured with a calliper to be 19.85 mm.

Relative is a measurement where the result is only a deviation from a pre-set value. For example, bolt height measured with a dial indicator (Figure 2.1) zeroed at 20.00 mm, showing the result/deviation in negative (−0.05 mm).

2.3 Measuring Instruments

2.3.1 *Types of instruments*

The variety of measuring instruments used in quality control is huge, but in general, they can be divided into two main groups: analogue and digital. The majority of analogue instruments are mechanical (see Figure 2.2) and do not require power, which makes them very reliable and independent. However, in the last few decades, digital instruments have become more and more popular because they are very versatile, offer more options in terms of units, less prone to misreading errors and can often be connected to a computer for automated data storage, processing, etc.

Nowadays, the variety of digital instruments is huge, ranging from a simple digital micrometre (Figure 2.3(a)) to three-coordinate measuring machines (3CMMs) (Figure 2.3(b)) that can do a lot more than just measure length, such as scanning, checking geometric tolerances, e.g. roundness, cylindricity, perpendicularity, flatness.

2.3.2 *Metrological characteristics*

The most important characteristics of the measuring instruments are accuracy and range. The **accuracy** of a measuring instrument is an essential part of the overall *measurement uncertainty* [4] and is usually described by its instrumental **error**, which is the difference between the *true value* (typically unknown but could be identified accurately enough if the measurement is done using a very high-precision instrument and is often called *actual value*) and the reading of the instrument.

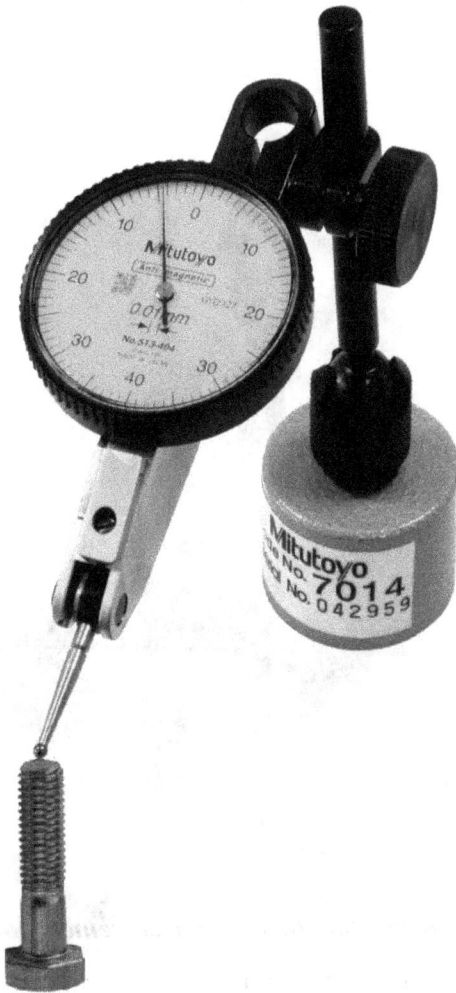

Figure 2.1.   Dial indicator used for relative measurements.

The measuring **range** is the difference between the highest and the lowest values that can be obtained by the instrument.

Usually, the metrological characteristics alongside other characteristics, such as *incremental value* (analogue instruments), *resolution* (sensitivity in digital instruments) and *repeatability* (variation in repeated measurement the same object) are presented in the specification document of the instrument.

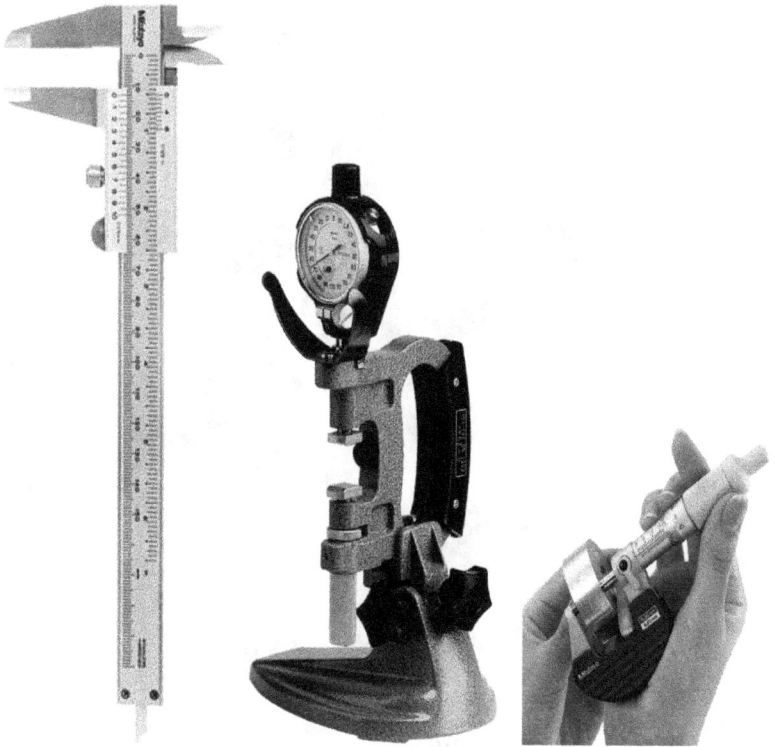

Figure 2.2.   Analogue instruments (from left to right): calliper, snap gauge and micrometre.

2.3.3 *Selection of instruments for measurement jobs*

The selection of an instrument for a job is mostly based on the above characteristics. A typical rule of thumb in respect to the accuracy is that the error of the instrument can be *no more than one-third to one-tenth of the specified tolerance of the measurand.* However, often in industry the selection is influenced by a wide range of characteristics such as:

- productivity, often defined by time per single measurement, which is important to avoid 'bottle necks';
- reliability, defined by the expected time to tool failure;

(a)

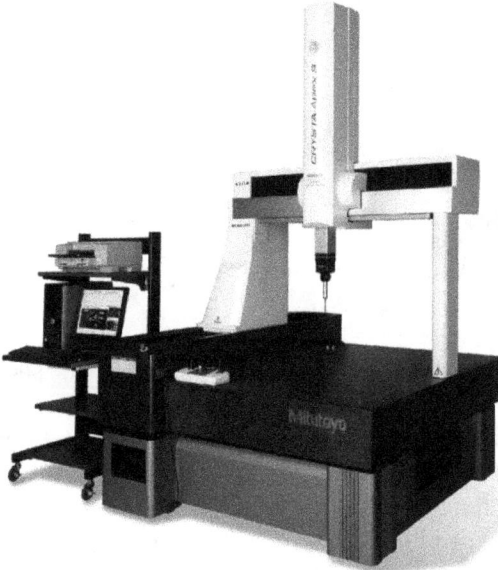

(b)

Figure 2.3.   (a) Digital micrometre; (b) 3CMM.

- cost, which affects the overall cost of the product;
- traceability (see following sections);
- availability.

2.4 Measurement Errors

By definition, *measurement error* is the difference between a measured value of a quantity and its true value. It is essential to understand that *no measurement is perfect*. Inevitably, all measurement results include some

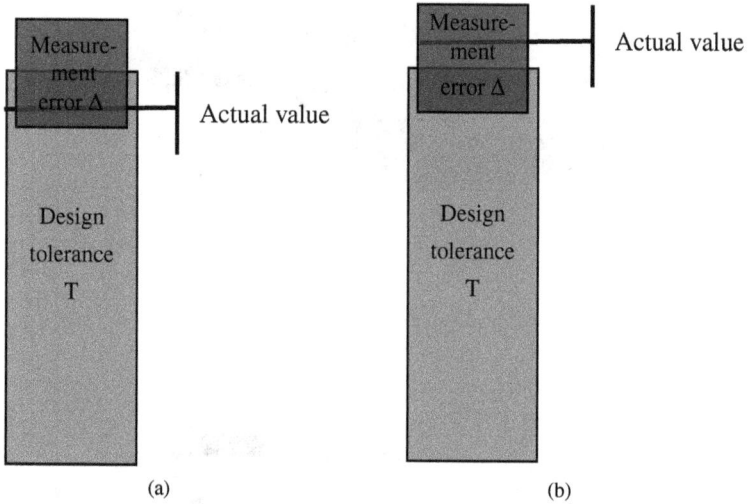

Figure 2.4.   Misjudgement errors: (a) error type I; (b) error type II.

uncertainty due to the combined effect of different errors, such as inherent instrumental error, operator's error and ambient conditions change. For this reason, a measurement result is complete only when it is accompanied by a statement of the associated uncertainty, e.g. standard deviation (see Chapter 4). In extreme situations, the measurement uncertainty may lead to misjudgements resulting in *rejecting a good part (error type I)* or *accepting a bad part (error type II)*. For example, if the actual value to be measured is inside the specified tolerance *T* but is close to the limit (see Figure 2.4(a)), the measurement uncertainty/error may cause a deviation and the result may be read as being just outside the tolerance, which will lead to the rejection of a good part (error type I). If the actual value to be measured is outside the specified tolerance *T* but is close to the limit (see Figure 2.4(b)), the measurement error may cause a deviation and the result may be read as being just inside the tolerance, which will lead to accepting a bad part (error type II).

2.5 Calibration of Measuring Instruments

Over time and with use, the accuracy of measuring instruments may deteriorate and instrumental errors increase. In order to maintain the uncertainty

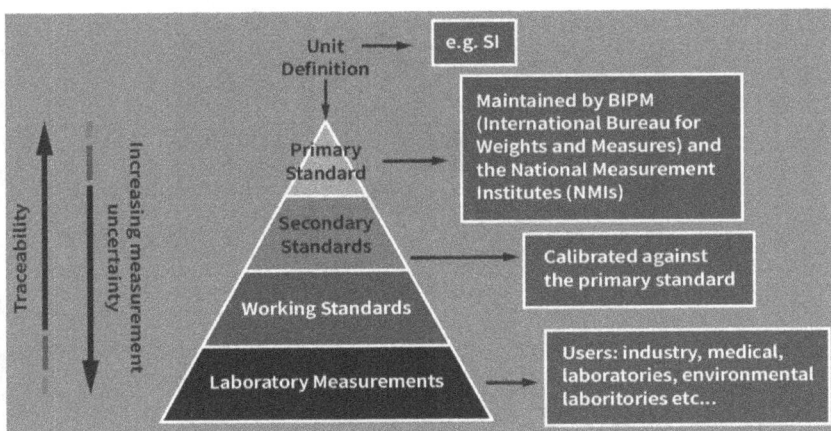

Figure 2.5.   Hierarchy of the traceability chain.

within an acceptable range, the measuring instruments are supposed to be calibrated. **Calibration** is a set of activities aimed to ensure that measuring instruments maintain their metrological characteristics. The instrument is compared to a higher-grade instrument or a *standard,* and adjustments are made if necessary. A calibration program includes inspection of the measuring instrument, maintaining a schedule and keeping records. Using calibrated instruments allows for **measurement traceability** [4], which is a property of the measurement result to be related to a primary reference standard through a documented unbroken chain of calibrations (Figure 2.5). To enable comparison of measurement results over time and distance, it is often beneficial for the reference to be a base unit of the International System of Units (SI).

So, for example, a ruler might be checked against a calliper, which can be checked against gauge blocks (working standard) (Figure 2.6) and so on.

Provided the required equipment is available, the calibration can be done in-house. Alternatively, it can be carried out by an accredited calibration laboratory which is supposed to have transfer standards checked against national reference standards (kept in a national measurement institute) which in turn are checked against the international/primary standard [5].

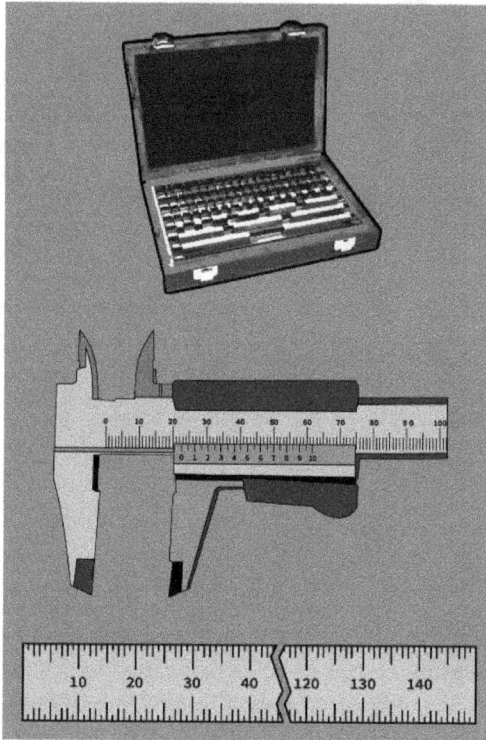

Figure 2.6.   Part of traceability chain involving a ruler (bottom of the chain), calliper and gauge blocks (top of the chain).

2.6 Limit Gauges

Often in mass production, where inspection should be fast and reliable, there is no need of measuring the actual size of a feature/product but only to check if it is within the tolerance or not. In such cases, very simple but highly productive tools, called **limit gauges**, are used.

2.6.1 *What is a limit gauge and how does it work?*

A limit (also known as GO/NOT-GO) gauge (see Figure 2.7) is an inspection tool that does not return a size in the conventional sense but instead returns a state. The state is either acceptable (the part is within tolerance and may be used) or unacceptable (and must be rejected).

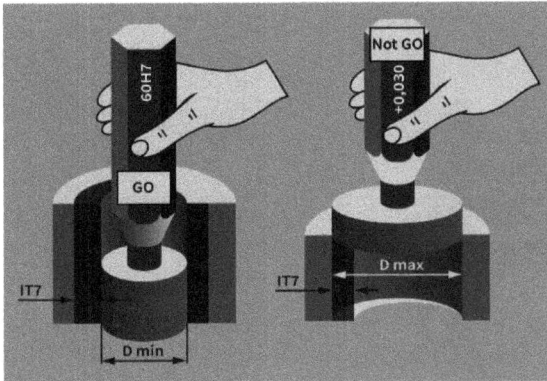

Figure 2.7. Plug limit gauge.

So, during inspection, if the GO end (the longer plug) fits in and the NOT-GO does not, as seen in Figure 2.7, then this hole will be accepted as good, i.e. the inspected diameter is within the tolerance. If both ends fit in, then this hole has been made too big and should be rejected. If both ends do not fit in, then the hole is too small and should be rejected but could also be classified as repairable. According to the *Taylor's principle of gauge design*, the GO end is designed to check the maximum material limit including form deviation (roundness, cylindricity), while the NOT-GO end is designed to check the minimum material limit and just one element of dimension at a time.

2.6.2 *Types of limit gauges*

There are different types of limit gauges.

- According to the application:
  The limit gauges are designed according to the purpose of their use. Figure 2.8 shows snap limit gauges that are used for inspection of a shaft's diameter. Often, they can be designed as two separate (GO and NOT-GO) rings. Similarly, limit gauges are developed for the inspection of a variety of features, such as threads, splines, keys, taper and round.
- According to the size—fixed or adjustable:
  The limit gauges are highly specialised instruments. Typically, a snap or plug limit gauge is designed to measure a specific dimension with a given tolerance. However, some of them can be made adjustable (Figure 2.9), allowing for a bit more dimensional flexibility.

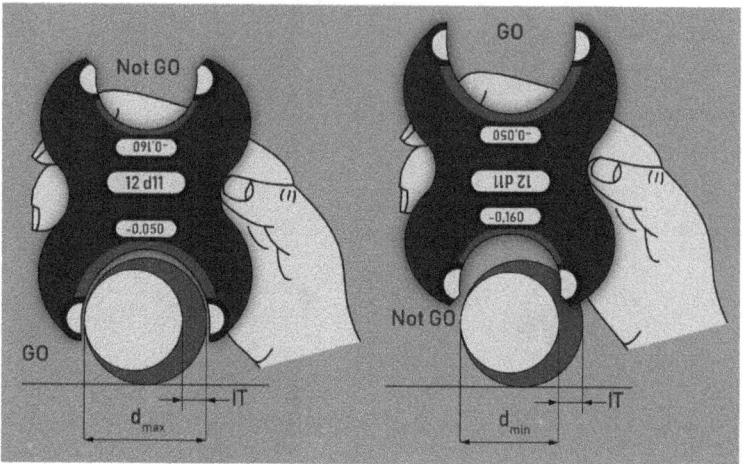

Figure 2.8. Two-sided (double) snap limit gauge.

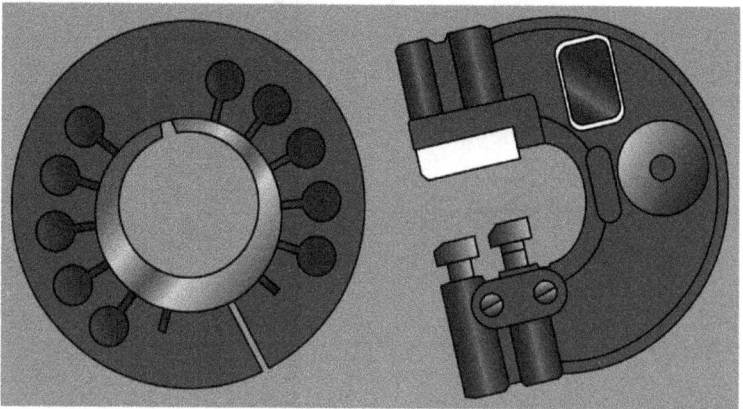

Figure 2.9. Adjustable thread ring gauge (left) and snap limit gauge (right).

2.6.3 *Design of limit gauges*

Generally, in order to guarantee stability, durability and precision, the main parts of the gauges are made of high-quality carbon steel with suitable heat treatment (usually hardness of at least 58 HRC) and a high degree of surface finish.

2.6.3.1 *Limit gauge tolerances*

Gauges, like any other jobs, require manufacturing tolerances due to inevitable imperfections in the workmanship of the gauge maker. So, the *gauge maker requires tolerances* to manufacture the GO and NOT-GO ends of the gauge. In order to maintain the accuracy of the inspection, the gauge tolerance should be kept as minimum as possible but too tight a tolerance would incur high costs to do so.

- The tolerance of the GO and NOT-GO gauges is usually 10% of the tolerance of the dimension to be inspected (workpiece tolerance).
- The tolerance of the ends of a limit gauge is approximately *10% of the tolerance to be checked.*
- The GO end in addition has some *wear allowance* (around 50% of the gauge tolerance).

2.6.3.2 *Limit gauge standardisation*

The limit gauge tolerances of plain plugs and rings have been standardised (see Table 2.1) according to the workpiece tolerance.

Table 2.1. Tolerances for plain limit gauges, mm (BS 969:2008).

Workpiece Tolerance, mm	Tolerance, T, for GO and NOT-GO Gauges, mm	Wear Allowance, W, for GO Gauges Only
$0.009 \leq 0.018$	0.001	0.001
$> 0.018 \leq 0.032$	0.002	0.001
$> 0.032 \leq 0.058$	0.003	0.002
$> 0.058 \leq 0.100$	0.004	0.004
$> 0.100 \leq 0.180$	0.006	0.007
$> 0.180 \leq 0.320$	0.009	0.012
$> 0.320 \leq 0.580$	0.014	0.025
$> 0.580 \leq 1.000$	0.025	0.048
$> 1.000 \leq 1.800$	0.040	0.080
$> 1.800 \leq 3.200$	0.050	0.155

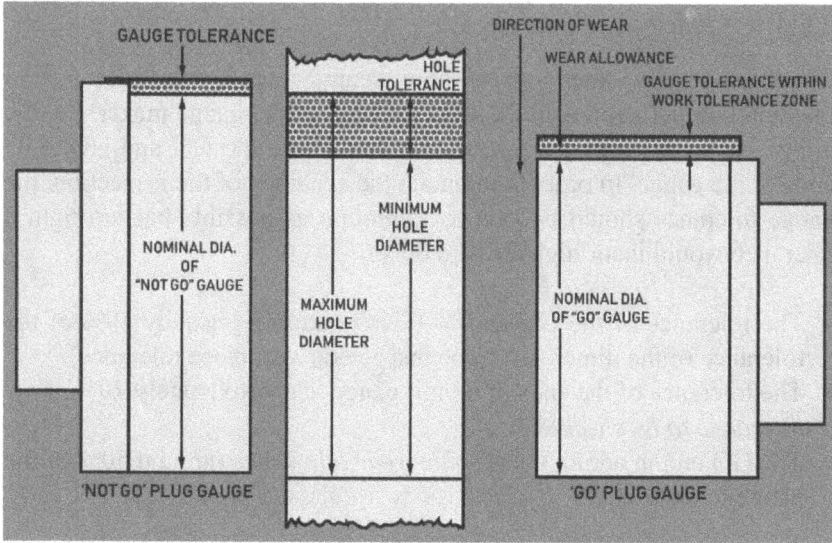

Figure 2.10. Tolerances of plug gauges.

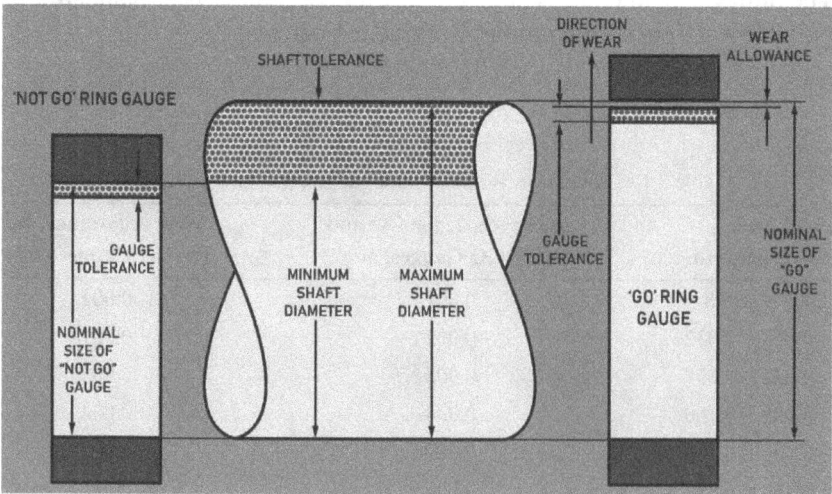

Figure 2.11. Tolerances of ring gauges.

A graphical representation of the magnitudes of tolerances and wear allowance (for the GO plug) of plug gauges in respect to the measured tolerance is shown in Figure 2.10. Similarly, Figure 2.11 shows tolerances and wear allowance (GO ring) for ring gauges. In both cases, wear allowance is allocated to the GO end only as the NOT-GO end is not supposed to wear.

2.6.3.3 *Limit gauge tolerance disposition*

Similar to the conventional measuring instruments (callipers, micrometres, etc.), inspection using limit gauges inevitably involve some risks of misjudgement—error types I and II. However, unlike the other instruments, while using limit gauges, these risks are manageable. The position of the gauge tolerances in respect to the workpiece tolerance is important as it defines a certain type of *risk of misjudgement*, which can be managed according to the company's quality policy. There are three distinct schemes for tolerance disposition.

Preventing error type II: this scheme (shown in green in Figure 2.12) has both tolerances and wear allowance fitted inside the workpiece tolerance. It is recommended by the British Standard (BS 969) as it completely

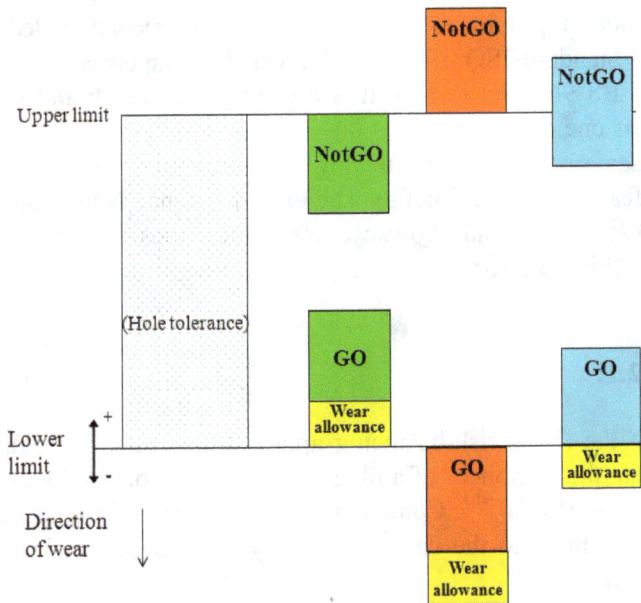

Figure 2.12. Tolerances of GO and NOT-GO plug gauges.

eliminates the risk of accepting a bad component as good. However, inevitably this scheme incurs an increased risk of rejecting good components (error type I). Often, to alleviate this risk and to prevent unnecessary waste, companies using this method double check the rejected parts with high-precision instruments.

Preventing error type I: this scheme (shown in orange in Figure 2.12) has both tolerances and wear allowance outside the workpiece tolerance so that it completely eliminates the risk of rejecting good components. This approach is not very popular among well-respected companies as it involves an increased risk for accepting bad components (error type II), which may lead to poor outgoing quality, customer dissatisfaction and ultimately may harm the image of the company. For this reason, this scheme is only used at workshop level but not for outgoing quality control.

Risk balancing (sometimes called 'Economical'): in this scheme, the NOT-GO gauge tolerance (in blue) is positioned half inside and half outside the workpiece tolerance; while the GO tolerance is located inside, the ware allowance is outside. Similar to the use of conventional measuring instruments, this approach benefits from reduced risks, but unlike the previous two, it does not eliminate any risks and it involves almost equal risks of both types of errors. This scheme is recommended by the European Standard ISO 1938-1-2015, which is supposed to (partially) supersede BS 969, but in fact, it is a different approach and should be regarded as one of the options.

The same schemes apply to ring and snap gauges when inspecting external features of size (shafts). The only difference is that the GO and NOT-GO tolerances and allowance are at the opposite workpiece tolerance limit (Figure 2.13).

Activity 2.1:

(a) Using the ISO's risk-balancing approach and data from Table 2.1, allocate the tolerances of a plug gauge for controlling the dimension of a hole of $\varnothing 30^{+0.1}_{0}$. Comment on possible risks in judgement and compare them to the case in which you use traditional measuring instruments.

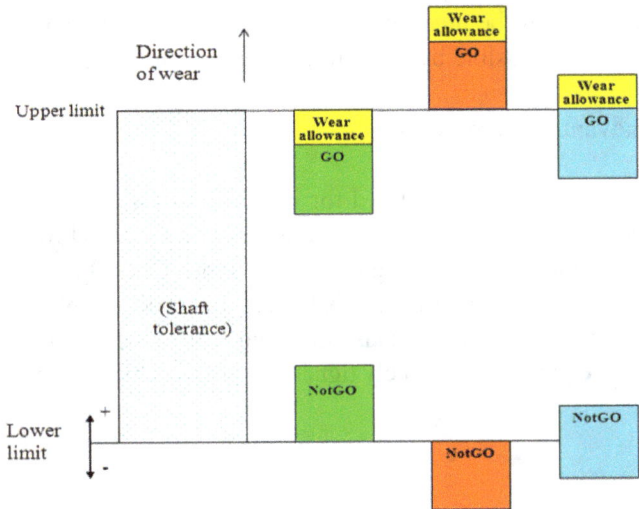

Figure 2.13. Tolerances of GO and NOT-GO ring/snap gauges.

(b) What would be the result of inspection of a hole having an actual diameter $\phi 29.998$ mm?

2.6.3.4 *Limit gauges use trade-off*

The main advantages of limit gauges use are as follows:

- They are very **cost-effective** in high-volume production because they offer a fast, simple (no need of highly qualified operators) and reliable inspection procedure.
- If the tolerances are properly designed, the limit gauges offer **manageable risks of misjudgement**. By selecting a tolerancing scheme, the designer can completely eliminate a certain type of error (I or II) in misjudgement.
- They can also check the **form deviation** apart from dimensional inspection.
- During inspection of inner threads, only the limit gauges offer **non-destructive testing**. Threaded holes are very hard to inspect as most of

the parameters to be measured (middle and outer diameters, pitch and angle) are inaccessible by conventional measuring instruments.

Some disadvantages are as follows:

- Most limit gauges are designed for inspection of a given specific size (nominal dimension and tolerance) and nothing else. However, some adjustable (more expensive) gauges offer more flexibility.
- Typically, the use of limit gauges is limited to inspection of tolerances no smaller than IT6. This is because the gauge tolerances for inspection of very precise parts are extremely tight and hard to achieve (see Table 2.1).

Chapter 3

Process Variation and Probability Distributions Used in Quality Management

3.1 Process Variation

Consider a simple operation, such as cutting steel bars, where a parameter (size) is monitored. This operation is part of a manufacturing process. Although the cutting machine has been set properly and the operation is repeated in the same way, a natural size variation around a certain value (target) will occur (Figure 3.1).

Generally, these variations may be due to a number of causes (e.g. power fluctuation, temperature changes, vibrations, tool wear, raw material inconsistency), any one of which will be of negligible effect. Yet, if most of them tend to the same direction, the size could steadily increase or decrease. It is, however, necessary to realise that **nothing is ever absolutely precise**.

3.1.1 *Tolerance*

As produced pieces always differ from one another, some margins must be set to limit the deviation from the target value. These margins are known as *tolerance* or *specification limits that form the range of the allowed error (deviation from the target)*. The term 'tolerance' is most commonly used for the dimensions of a manufactured item but can be used in any process. Thus, for example, the specification limits for the time a telesales operator may take to answer a customer call may be

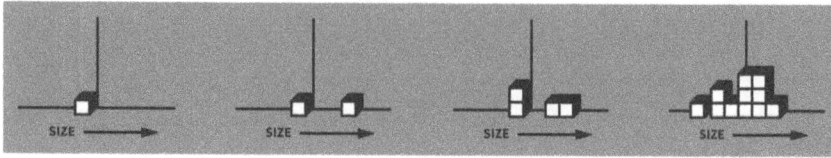

Figure 3.1.   Natural process variation.

between 0 and 5 s or the weight of a chocolate cake may vary between 290 g and 310 g and this still could be considered acceptable. At this point, there may be a conflict between the designer and the manufacturer. The designer will be tempted to 'play safe' by requesting the smallest tolerance he/she thinks will be needed to guarantee the quality, while the manufacturer, knowing that processes vary, will be anxious to have as large a margin for error as possible. The tolerance must be small enough to ensure acceptable and reliable performance of the product, but too tight a tolerance can result in considerable unnecessary expense due to the excessive amount of products scraped for being out of the tolerance. Of course, even a very simple product (e.g. a brick) usually has not just one but several tolerances (e.g. size, weight, strength) in the specification document, and they must all be met in order to be accepted as good. Anyway, in this book, we will not be dealing with the question of what tolerance to choose, which is the designer's job, but *how to make sure the product attributes (or characteristics) are within the required tolerances,* which is the quality manager's job. In order to be able to do this process, variation must be analysed.

3.2 Probability Distributions

If the production process is stable, the recorded classified values of the monitored parameter tend to form a variation **pattern** (Figure 3.2).

This pattern can be described as a **distribution** around a target value. Distributions can differ in location, spread and shape or any combination of these (Figure 3.3).

3.2.1 *Type of distributions*

There are two basic types of distribution: discrete and continuous. A *discrete distribution* is made up of elementary events that can only have

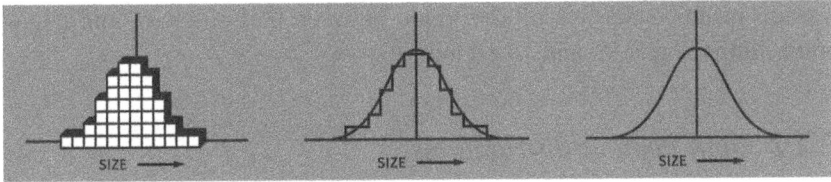

Figure 3.2.   Bell-shape variation pattern.

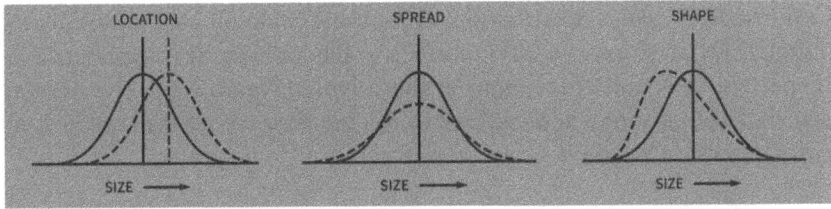

Figure 3.3.   Distribution with different locations (left), spread (middle) and shape (right).

Figure 3.4.   Example for generating a discrete distribution by repeatedly rolling a pair of dice.

distinct integer values. For example, if we roll a pair of dice (Figure 3.4), the numbers that turn up can sum up to any whole number from 2 to 12 and no others, i.e. it is not possible for them to sum up, say, to 4.5.

In the *continuous distribution,* the events may have any value between the given limits. For example, the wrench size of a nut produced by a certain machine may be of any value between the minimum and maximum limits, e.g. 9.90 and 10.10 mm.

3.2.2 *Some popular distributions*

3.2.2.1 *Normal distribution*

The normal, also known as Gaussian (named after the German scientist, Carl Friedrich Gauss) distribution, is the most common continuous distribution relevant to around 90% of cases in the industry. It is symmetric in shape with a distinct maximum at the middle (Figure 3.5). The normal distribution has many applications in the industry, so we will study it in more detail in Chapter 4.

3.2.2.2 *Exponential distribution*

It is a non-symmetric continuous distribution (Figure 3.6). This distribution is often used to model failure rate, tool life and reliability of production systems.

3.2.2.3 *Uniform distribution*

This is another continuous distribution, having a simple rectangular shape pattern (Figure 3.7). A typical example of application is with the lottery numbers (e.g. 1–49) probability, where every week, each number should have an equal probability of 1/48 (where 48 = 49 − 1) to occur.

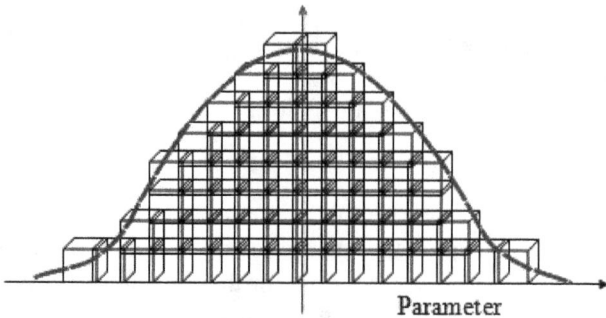

Figure 3.5.   Normal distribution.

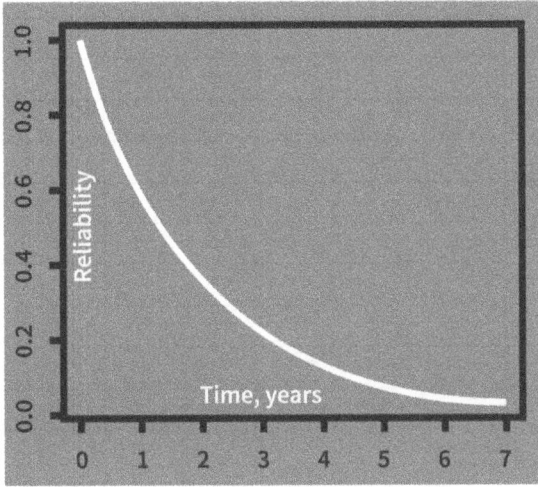

Figure 3.6.   Exponential distribution, time vs. reliability.

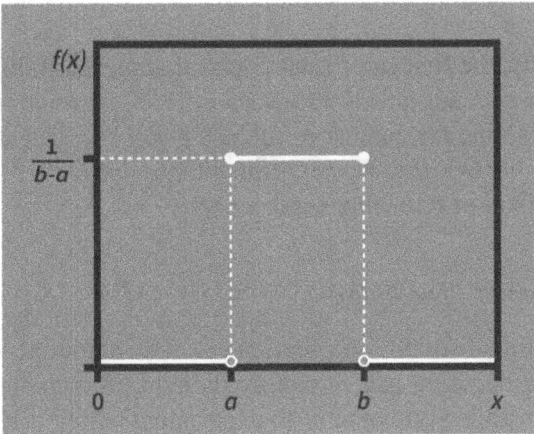

Figure 3.7.   Uniform distribution.

3.2.2.4 *Poisson distribution*

The Poisson distribution, named after the French mathematician, Simeon Denis Poisson, is a discrete probability. It is also known as low probability distribution as it is usually less than 0.10 in each trial. It is applicable to study and model some infrequent events, e.g. meteorites hitting in the next 100 years or the number of defective items in a batch of 100 (Figure 3.8).

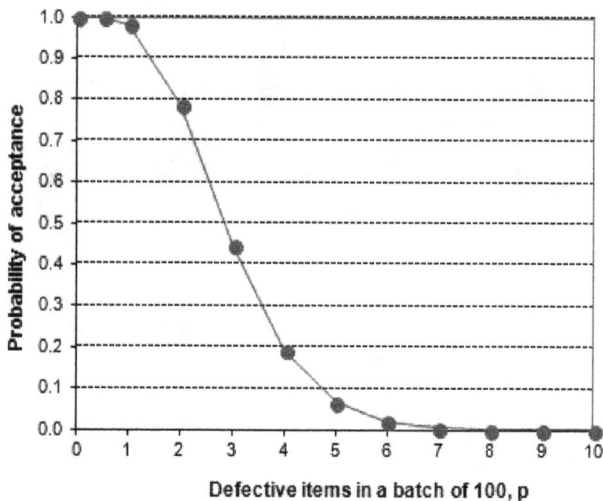

Figure 3.8.   Poisson distribution.

Historically, the first significant practical application of this distribution was made by Ladislaus Bortkiewicz in 1898, when he was given the task of investigating the number of soldiers in the Prussian army accidentally killed by horse kicks. This experiment introduced the Poisson distribution to the field of reliability engineering.

3.2.2.5 *Binomial distribution*

Another useful discrete distribution is the binomial distribution. A binomial distribution can be simply thought of as the probability of a successful or failed outcome in an experiment which is repeated multiple times. The binomial is a type of distribution that has two possible outcomes (the prefix 'bi' means two or twice). For example, a coin toss has only two possible outcomes: heads or tails and inspecting a batch of parts could have two possible outcomes: pass or fail (accept or reject).

The pattern of the binomial distribution (Figure 3.9) closely resembles the shape of the normal distribution, but the former is a discrete distribution as it deals with the integer number of occurrences unlike the latter.

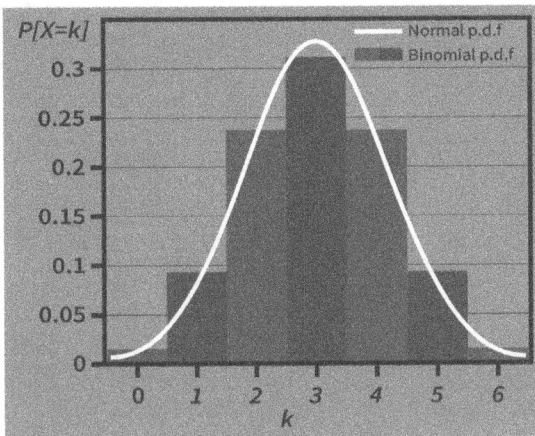

Figure 3.9. Binomial distribution.

This distribution is widely used in acceptance sampling (see Chapter 9).

3.3 Analysis of a Distribution

In order to extract the important information from a distribution, the data should be appropriately organised, correctly processed and carefully analysed. Simply looking at the number from measurements (even if they are presented in a table/text in a sensible way) may not be enough to make the most of the data. One of the most convenient ways to organise data for analysis is to classify the numbers and set up the so-called 'tally chart'.

Figure 3.10 shows the data from 100 observations classified based on the hardness (HRC) of a metallic material. The tally chart is a simple graphical tool, which can be used to analyse certain characteristics of the process variation and in particular how the data frequency is distributed along their value.

3.3.1 *Histogram*

A more modern way to visualise the distribution is by a histogram, where the height/length of each bar represents the frequency distribution (Figure 3.11).

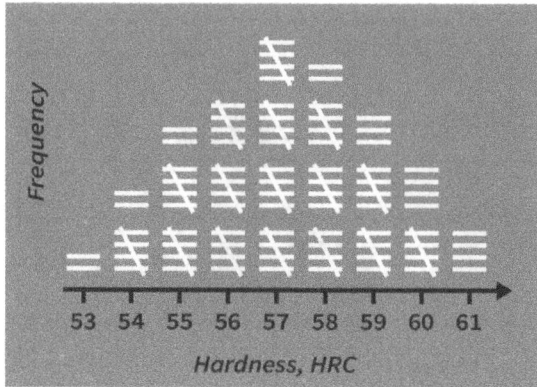

Figure 3.10.   Tally chart.

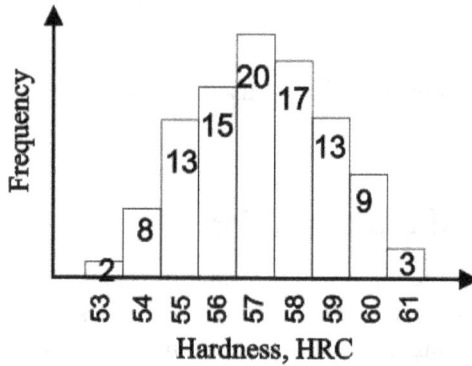

Figure 3.11.   Histogram.

3.3.1.1 *How to prepare a histogram?*

(1) Count the number of observations (in this case, there are 100).
(2) Identify the minimum X_{min} and maximum X_{max} values (in this case, 53 and 61, respectively).
(3) Calculate the range of the variation distribution R, $R = X_{max} - X_{min}$.
(4) Usually, the observations are grouped into classes/intervals. Typically, the number of classes K (the number of bars in the histogram) depends on the number of observations N:

$$K = 1 + 3.2 \log_{10} N, \tag{3.1}$$

where the result for K is rounded off to the closest whole number.

Table 3.1.   Recommended number of classes.

Number of Observations N	Recommended Number of Classes K
Under 50	5–7
50–100	6–9
100–250	7–12
Over 250	10–20

Often, to divide the range R into classes/intervals, the number of classes could be selected from a table, instead of the above formula, as shown in Table 3.1 (in this case, 9 is selected).

(5) Determine the class interval (represented by the width of the bar) d:

$$d = (X_{max} - X_{min})/K. \tag{3.2}$$

(In this case, $d = 1$ unit HRC, but it could be any non-integer number.)

(6) Plot the tally chart, counting the observations falling in each interval.

(7) On the basis of the tally chart, plot a histogram using an appropriate scaling factor for the bar height.

3.3.1.2 *What can we use a histogram for?*

The histogram is a very powerful tool used in QC, especially if having to dealing with a large number of observations. Looking at a properly built histogram, we can acquire some very important information.

We may try *guessing the nature of the distribution*. If we connect the middle points of the top end of each bar, the shape of the curve/polyline (see dashed line in Figure 3.12) may help us to identify the distribution. In this case, where there is central symmetry around a single maximum, our best guess would be a normal distribution.

We could also try to *predict the next value* that would occur in the future. Observing that the highest frequency is around 57, we may say that it is quite likely that the next value would be around this value. This way, we turn the *frequency distribution* that the histogram represents into a *probability distribution*.

If we add tolerance limits to the histogram (see Figure 3.13), we could try to *estimate the percentage of the scrap*. In this case, as we have a total

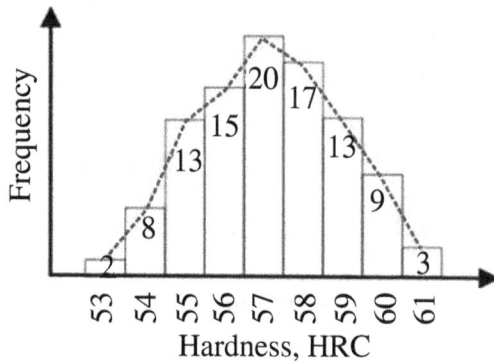

Figure 3.12. Histogram with empirical curve/polyline.

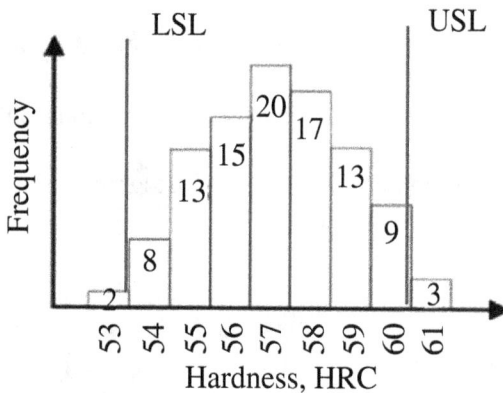

Figure 3.13. Histogram.

of five items from 100 observations out of the tolerance, we could make a rough estimate that we would expect 5% scrap.

3.3.2 *Some typical cases in distribution analysis using histograms*

Often, when analysing the distribution through the shape of a histogram, we may observe a pattern similar to the one in Figure 3.14. Unlike the classic normal distribution, there is more than one maximum in this case. Usually, this is an indication of a mixed dataset, e.g. data from two

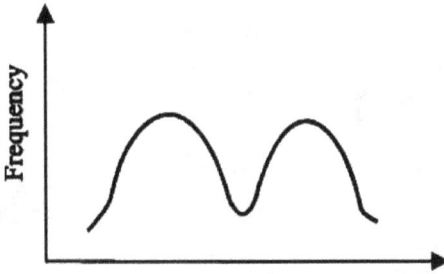

Figure 3.14.   Bimodal distribution.

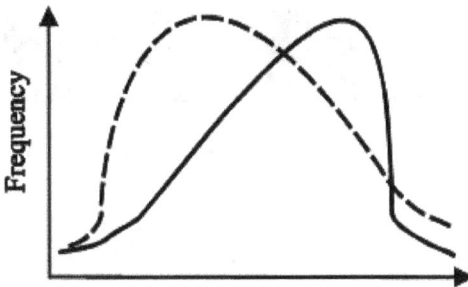

Figure 3.15.   Skewed distribution.

different machines and one machine with two different process settings. Ideally, this problem should be solved by filtering/segregating the data.

Another typical case is shown in Figure 3.15. Obviously, this distribution lacks the central symmetry around the maximum. Usually, this is the result of the influence of a systematically dominant changing factor that is shifting the process setting to a certain direction, i.e. up or down. In manufacturing, a common cause for this anomaly is tool wear.

3.4 Normal Distribution

Theoretically, a random variable x has a normal distribution, which is often denoted by $N(\mu, \sigma^2)$ if it has a *probability density function (PDF)* as follows:

$$f(x, \mu, \sigma^2) = \frac{1}{\sigma \sqrt{2\pi}} e^{-\frac{(x-\mu)^2}{2\sigma^2}}. \tag{3.3}$$

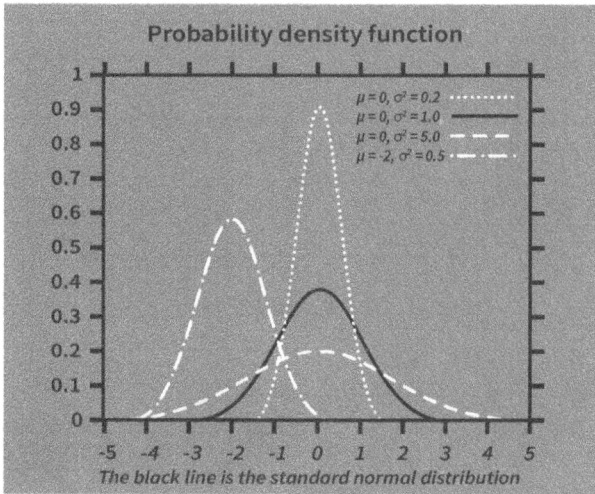

Figure 3.16. Probability density function plots for different mean μ and standard distributions σ.

This function has two parameters (mean μ and standard deviation σ) and if we plot it, we would get the *normal distribution curve*, see Figure 3.16.

3.4.1 *Why is the normal distribution so important?*

The normal distribution curve has a very useful property. The area under the curve represents the probability, that the event '*the random value x to fall within the certain limit*' will occur. For example, in Figure 3.17, the shaded area as a portion of the whole area represents the probability that the random variable x will fall in the interval ($-\infty$ to x_1). The total area under the curve is one (or 100%), which analytically can be expressed as follows:

$$\sum_{i=-\infty}^{\infty} P(x_i) = \int_{-\infty}^{\infty} f(x)dx = 1. \qquad (3.4)$$

Then, the shaded area under the curve (hence the probability) can be calculated between various points along the abscissa:

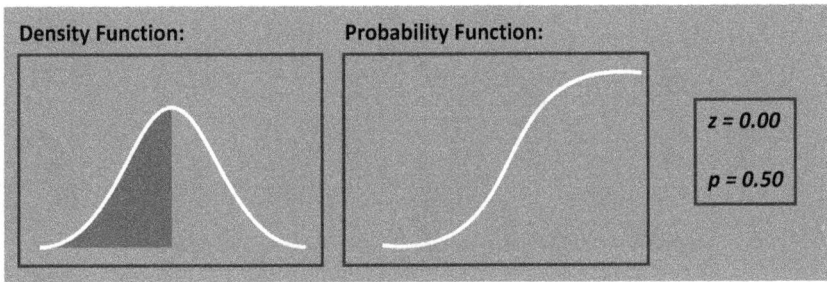

Figure 3.17.   Shaded area represents 50% probability.

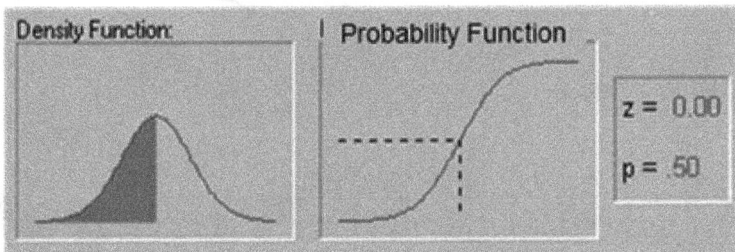

Figure 3.18.   Density function transformed to probability/cumulative function.

$$\sum_{i=-\infty}^{x_1} P(x_i) = \int_{-\infty}^{x_1} f(x)\,dx. \qquad (3.5)$$

So, for a density function with a given mean (μ) and standard deviation (σ) for each **x**, the probability is calculated transforming it into a *probability function*, also known as a *cumulative distribution function*. If plotted, the latter will have an asymmetric shape (see Figure 3.18) ranging from 0 to 1 (0–100%).

3.4.2 *Standard (unit) normal distribution*

In the general case, calculations using the formula in Equation (3.5) are complicated. To facilitate them, a special *standard* (also known as *unit*) *normal distribution* is introduced. The standard normal distribution has mean $\mu = 0$ and standard deviation $\sigma = 1$ (see Figure 3.19) and is usually

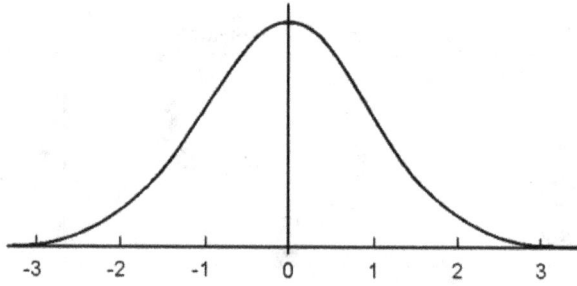

Figure 3.19. Standard normal distribution.

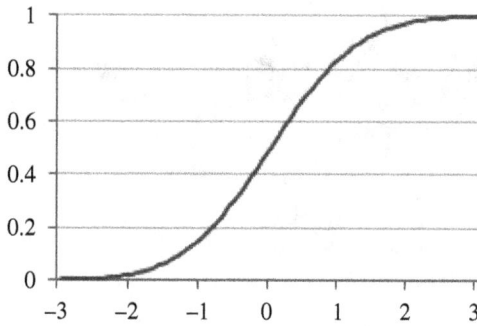

Figure 3.20. Standard normal cumulative distribution.

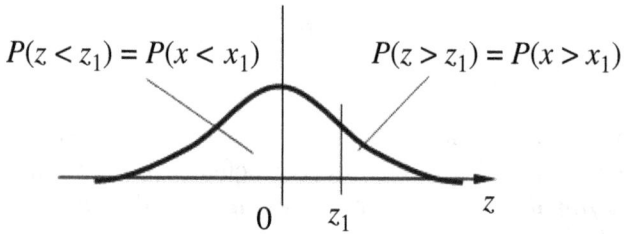

Figure 3.21. Probability $P(X) = P(Z)$.

denoted by $N(0,1)$. The standard normal distribution is much easier to work with and to calculate the area under the curve, which is tabulated, i.e. to obtain the **standard normal cumulative distribution** (Figure 3.20).

The standard normal (cumulative) distribution function of z, denoted by $\Phi(z)$, defines the area (probability, P) under the curve in exactly the same way as for x (Figure 3.21).

So, the area under the curve in $(-\infty, z_1)$ is equal to the area in $(-\infty, x_1)$ and is defined by

$$\Phi(z) \equiv P(x \le x_1) \equiv P(z \le z_1)$$

$$= P\left(z \le \frac{x_1 - \mu}{\sigma}\right) = \frac{1}{\sqrt{2\pi}} \int_{-\infty}^{z_1} e^{\frac{-z^2}{2}} dz. \tag{3.6}$$

The values of $\Phi(z)$ are tabulated (also known as the Z-score table) for the range $-4 < z < 4$ (actually $\pm 4\sigma$), see Appendix 3.1. To find the $\Phi(z)$ values in Excel, use NORMSDIST (z) function.

3.4.3 *How to transform a normal distribution to a standard (unit) normal distribution?*

Any normal distribution can be converted into a standard (unit) normal distribution by the so-called *Z transformation* using the formula

$$z = \frac{x - \mu}{\sigma}. \tag{3.7}$$

The new variable, z, represents the number of standard deviations (σ) that the random variable x is away from the mean (μ). In this sense, z can be considered as a unit to measure the distance from the distribution centre, see Figure 3.22.

Example 3.1: In a manufacturing process, the time required to produce a certain electronic component is normally distributed with a mean of 50 s

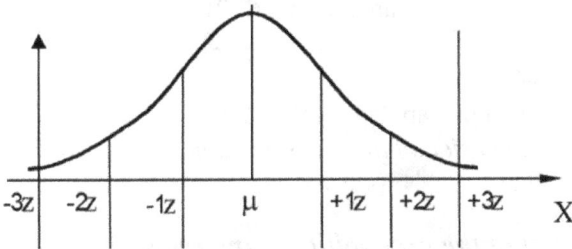

Figure 3.22. Z used as a unit to measure the distance from the distribution centre.

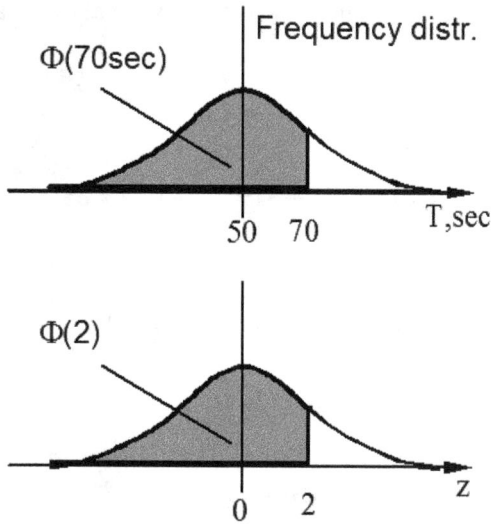

Figure 3.23.   Probability distribution before (top) and after (bottom) the Z transformation.

and a standard deviation of 10 s (Figure 3.23). What is the probability that the process will take less than 70 s?

Answer:
We have $N(50, 10)$.
   Our targeted value is time $(T) = 70$ s.

$$z = \frac{T - \mu}{\sigma} = \frac{70 - 50}{10} = 2.$$

From the cumulative distribution tables,

$$\text{area } \Phi(z) = \Phi(2) =$$
$$= 0.97725 = 97.725\%.$$

Please note that the tabulated value $\Phi(z)$ represents the area (probability) on the *left-hand side of the targeted value.*

3.4.4 *Evaluating the probability in intervals*

Given that in the industry we work with tolerances, we are often interested in estimating the probability of two targeted values (x_1 and x_2) forming a

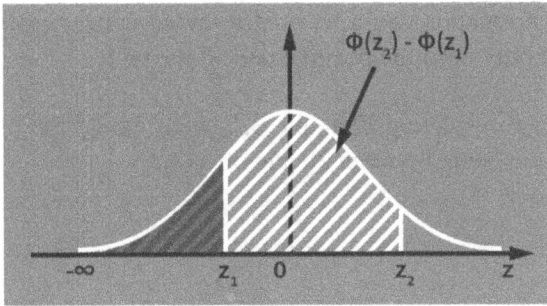

Figure 3.24.   Probability in the interval (z_1, z_2).

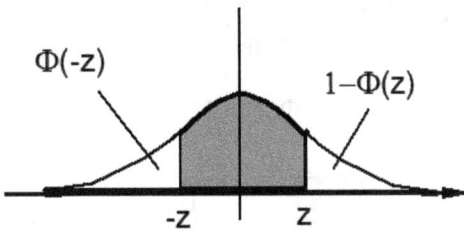

Figure 3.25.   Central symmetry of the normal distribution density function.

closed interval, i.e. a tolerance. Analytically, this can be achieved using the following equation:

$$P(x_1 < x < x_2) = \frac{1}{\sqrt{2\pi}} \int_{z_1}^{z_2} e^{\frac{-z^2}{2}} dz = \Phi(z_2) - \Phi(z_1). \qquad (3.8)$$

Practically, all we have to do to solve this task is to simply apply the approach from Example 3.1 twice. First, we need to do the Z transformation twice to find the z_1 and z_2 corresponding to x_1 and x_2. Then, using the table, we can find the areas (probabilities) $\Phi(z_1)$ and $\Phi(z_2)$. As these probabilities represent the area under the curve from $-\infty$ to z_1 (dark hatched area, Figure 3.24) and from $-\infty$ to z_2 (bright hatched area), respectively, in order to find the area (probability) in the interval (z_1, z_2), we have to subtract $\Phi(z_1)$ from $\Phi(z_2)$.

Hint: It might be useful to note that from the central symmetry of the normal distribution density function of z about 0, illustrated in Figure 3.25, it is apparent that

$$\Phi(-z) = 1 - \Phi(z). \qquad (3.9)$$

This transformation could be used to avoid using negative z values, i.e. we would only need the positive half of the table.

Example 3.2: Considering Example 3.1, what would the probability of the process time being between 30 and 50 s be?

Answer:

$$z_1 = \frac{T_1 - \mu}{\sigma} = \frac{30 - 50}{10} = -2.$$

Hence, from the table, $\Phi(z_1) = \Phi(-2) = 0.02275$.
Observing (3.9), we could obtain the same by

$$\Phi(z_1) = \Phi(-2) = 1 - \Phi(2) = 1 - 0.97725 = 0.02275,$$

$$z_2 = \frac{T_2 - \mu}{\sigma} = \frac{50 - 50}{10} = 0.$$

Hence, from the table, $\Phi(z_2) = \Phi(0) = 0.5$.

Finally (Figure 3.26), $P(30 < T < 50) = \Phi(z_2) - \Phi(z_1) = 0.5 - 0.02275$
$$= 0.47725 = 47.725\%.$$

Activity 3.1: A random sample of 100 machined shafts produced by a manufacturing firm was taken, and their diameters were measured with the following results: mean = 50.05 mm and standard deviation = 0.1 mm. If the tolerance specified for the process is 50 ± 0.15 mm, what is the

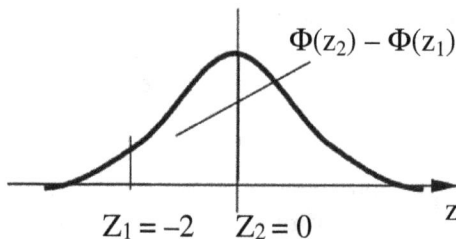

Figure 3.26. Probability $P(30 < T < 50)$.

expected percentage/number of rejects in a batch of one million? Support your answer with relevant sketches.

Activity 3.2: If in the 100-shafts sample in Activity 3.1, there were only five observations out of the tolerance 50 ± 0.15 mm, does this imply that 5% were rejects? If yes, compared with the result from Exercise 3.1, which estimate is more reliable and why?

Chapter 4

Sampling

4.1 Sampling

When we deal with large quantities from high volume production, we often physically do not have the time to observe/test every single produced item. In such cases, we may consider sampling.

Usually, the term 'sampling' involves the act of taking samples. We have already used the term 'sample', but perhaps we still need to clarify its meaning. What exactly is a sample? By definition, it is *a portion of a population or an object selected and taken so that the observed value(s) has (have) no bias.*

The meaning of 'sample' should not be mixed up with the term 'specimen'. Usually, in the industry, when dealing with high volume production, the sample we take to observe/inspect frequently consists of more than one item/specimen.

A very important property of the sample is to be representative for the entire population (lot) or object. Often, with batch production making the sample representative (no bias), the observed items are randomly selected. In continuous/automated production, samples can be taken on a regular basis, e.g. every hour on the hour, every shift, every day. If the production is intermittent (by order), this could be done, for example every batch or every 100 parts produced.

4.1.1 *Sample characteristics*

4.1.1.1 *Sample size*

Usually, samples with 1 to 10 items/observations are considered *small*. Sample sizes from 10 to 100 are regarded as *medium*, while those over 100 are *big*.

Assume that we have a sample of observations from a normally distributed population. If plotted, the data are expected to look like that in Figure 4.1. The peak of the curve shows the sample data *location*, typically associated with the setting of the process. The spread of the curve represents the data *variation* range, which is associated with the accuracy/precision of the process.

For example, the plot in Figure 4.2 represents the data from a machine that was run with two different settings.

Figure 4.3 represents sample data plots from two machines having the same settings but different accuracies.

To estimate these very important sample characteristics (location and variation), we need some numerical measures. These measures are also known as *descriptive statistics*.

4.1.1.2 *Sample measures*

(1) **Measures of location (central tendency):**
(a) The most popular measure of data location is the *sample mean (average).*

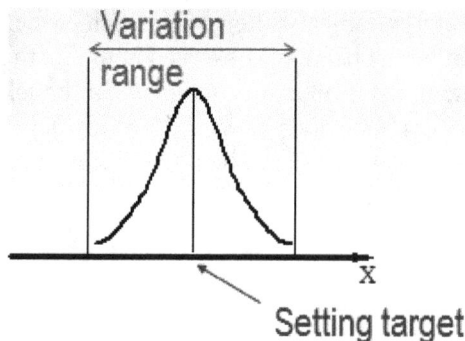

Figure 4.1.   Sample distribution.

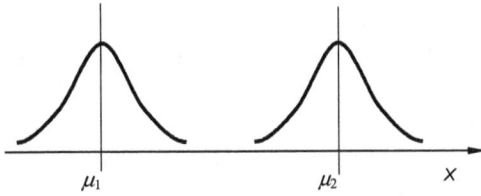

Figure 4.2.   Sample distributions from two machines/processes having different settings.

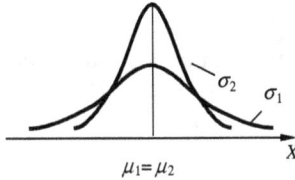

Figure 4.3.   Sample distributions from two machines/processes having the same settings, but different accuracies.

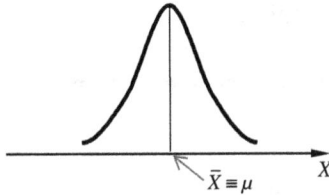

Figure 4.4.   Sample mean representing the population mean.

The sample mean $\bar{X}$ (pronounced X bar) is an estimate of the population mean μ. If the sample is representative and large enough, the sample mean (average) is supposed to be very close to the population mean μ, i.e. $\bar{X} \equiv \mu$, see Figure 4.4.

The average is calculated using the formula

$$\bar{X} = \frac{1}{n}\left(X_1 + X_2 + ... + X_n\right) = \frac{1}{n}\sum X_i. \tag{4.1}$$

Note: It is recommended that when calculating the average $\bar{X}$ value, it is rounded off to one extra decimal position.

(b) Median
The median $\tilde{X}$ (X wave) is an estimate of the location of an ordered sequence of data/ordered dataset. It represents the middle of the dataset,

i.e. it is the number that splits the dataset into two equally sized halves, one half containing the numbers that are smaller than the median and the second half containing numbers bigger than the median. For example, if we have a sequence of five numbers in ascending order:

1, 2, 3, 6 and 8, then the median will be the member located in the middle of the sequence (3).

$$\tilde{X} = X_{\frac{n+1}{2}} \quad \text{if number of data } n \text{ is odd.} \tag{4.2}$$

In a dataset, with an even number of members, the median is equal to the average of the middle two members. So, in the sequence:

4, 4, 5, 10, the median $\tilde{X} = 4.5$.

$$\tilde{X} = \frac{X_{\frac{n}{2}} + X_{\frac{n}{2}+1}}{2} \quad \text{if } n \text{ is even.} \tag{4.3}$$

A large difference between the mean and the median usually is an indication of a significant skew in the data, which might be due to lack of normality or presence of outliers (see Section 4.2.3).

(c) Mode

The mode (Mo, as in 'most often') is used for datasets where there are repeated/grouped members. Mo is equal to the value (not the frequency) of the member that occurs most often. For example, in Figure 3.10, the mode is 57.

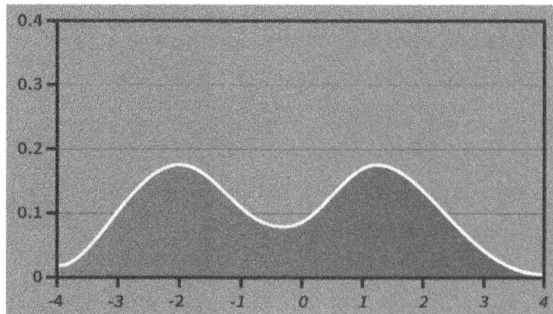

Figure 4.5. Bimodal (−2 and 2) dataset.

It is not uncommon for a dataset to have more than one mode. This happens when two or more members occur with equal frequency in the dataset. A dataset with two modes is called bimodal, see Figure 4.5.

A dataset with three modes is called tri-modal and so on.

Usually, the presence of more than one mode indicates poor homogeneity of data, e.g. data taken from different processes, machines, etc.

Often, when estimating a set of data, the three location measures may show different results, see Figure 4.6. Let us compare the three measures of sample location in terms of accuracy and practicality. What can we say? Analytically, the average is perhaps the best measure as it takes all the members in the dataset into account, while for example the median is only the middle-standing member of sequence whatever this is. This may be a bit misleading while judging the tendency. From this point of view, if we have repeated observations, the mode might be useful, but generally it would require a significant amount of data.

(2) **Measures of spread (variation):**

Let us compare two sample datasets: A (4, 5, 5, 5, 6, 6, 6, 6, 7, 7, 7, 8) and B (1, 2, 3, 4, 5, 6, 6, 7, 8, 9, 10, 11). Note that each dataset has all three measures of central tendency/location equal to 6. If we just looked at the measures of central tendency, we may assume that the two datasets are the same. However, if we look at the spread of the values in the graph

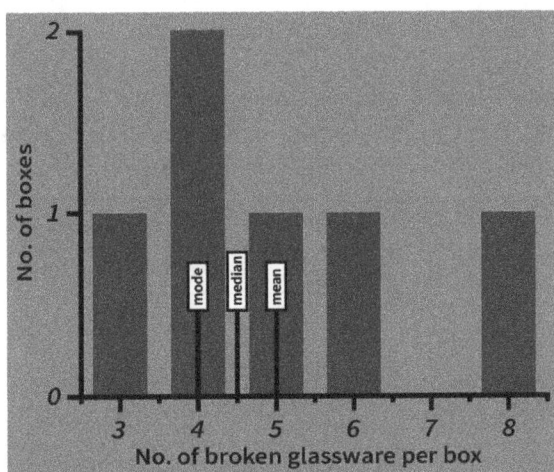

Figure 4.6. Mode, median and mean of a data set may have slightly different values.

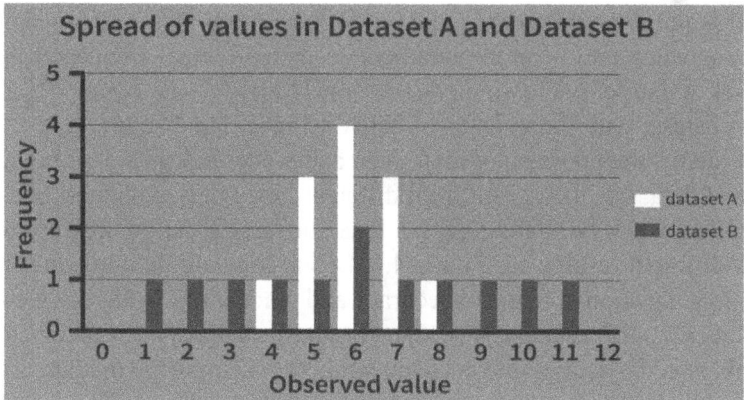

Figure 4.7.   Two data sets (A and B) distributions.

(Figure 4.7), we can see that dataset B is more dispersed than dataset A. Used together, the measures of central tendency and the measures of spread help us to understand the data better.

(a) The simplest measure of spread is the *range R*, which is calculated as the difference between the smallest value and the largest value in a dataset:

$$R = X_{max} - X_{min}. \tag{4.4}$$

Dataset A will have $R = 4$, while dataset B will have $R = 10$, indicating a significant difference between the two.

(b) A more sophisticated measure of the spread is the *variance* σ^2. The population variance σ^2 (pronounced sigma squared) of a discrete set of numbers is equal to the averaged squared deviation of each observation from the mean, μ. It is expressed by the following formula:

$$\sigma^2 = \frac{\sum_{i=1}^{N}(X_i - \mu)^2}{N}, \tag{4.5}$$

where
X_i represents each observation, starting from the first to the last,
μ represents the population mean and
N represents the number of observations in the population.

Please note that on averaging (dividing by N), the deviations allow us to compare the spreads of different-sized datasets.

The *variance of a sample,* S^2 (pronounced s squared) is expressed by a slightly different formula:

$$S^2 = \frac{\sum_{i=1}^{n}(X_i - \bar{X})^2}{n-1},\qquad(4.6)$$

where
$\bar{X}$ represents the sample mean and
n represents the number of observations in the sample.

(c) People often get confused by the units of variance, which is squared, e.g. kg^2, V^2 and min^2. For this reason, a more popular measure of spread is used called the *standard deviation* σ.

The standard deviation σ (often denoted by SD) is just the square root of the variance:

$$\sigma = \sqrt{\frac{\sum_{i=1}^{N}(X_i - \mu)^2}{N}}.\qquad(4.7)$$

Please note that the SD for a population is usually represented by σ, and the SD for a sample is represented by S:

$$S = \sqrt{\frac{\sum_{i=1}^{N}(X_i - \bar{X})^2}{n-1}}.\qquad(4.8)$$

For sample (dataset) A, the variance $S^2 = 1.27$ and the SD $S = 1.13$. For sample (dataset) B, the variance $S^2 = 10.00$ and the SD $S = 3.16$. The larger variance and standard deviation in dataset B further demonstrates that dataset B is more dispersed than dataset A.

Because it is so popular, often SD (σ) is used as a unit to measure the process data spread (variation) interval. For example, it is known that the interval $\pm 1\sigma$ represents 68% of the entire population, see Figure 4.8. The interval $\pm 2\sigma$ (often used in metrology) represents 95% of the entire population. The range $\pm 3\sigma$ represents 99.73% of the entire population, and it is often called *process variation* (V) or *natural tolerance interval.*

(d) Another powerful measure of spread is the root-mean-square deviation (RMSD) or root-mean-square error (RMSE) (or sometimes root-mean-squared error). The RMSD is defined as the square root of the mean-squared error:

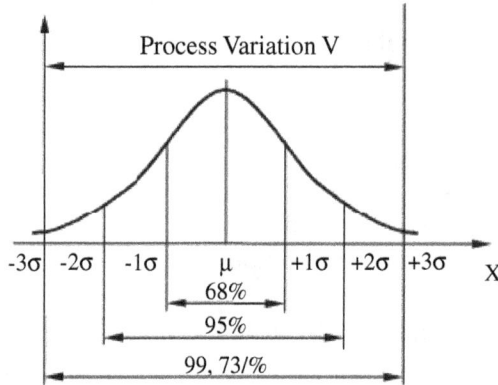

Figure 4.8.   SD (σ) used as a unit to measure the process data spread (variation) interval.

$$RMSD = \sqrt{\frac{\sum_{i=1}^{N}(X_i - X_i^{nom})^2}{n}}. \qquad (4.9)$$

It is a frequently used measure of the differences between two equal-sized datasets (sample or population values), where one set is values *predicted* by a model or an estimator/nominal (X_i^{nom}) and the other set is the *observed* values X_i.

Applying (4.9) to datasets A and B we get

$$RMSD = \sqrt{\frac{\sum_{i=1}^{N}(A_i - B_i)^2}{n}} = 2.$$

A simplified variation of the RMSD used when we measure the spread of a dataset against a single nominal value X^{nom} is

$$RMSD = \sqrt{\frac{\sum_{i=1}^{N}(X_i - X^{nom})^2}{n}}. \qquad (4.10)$$

In this case, the formula (4.10) resembles the original formula for SD (4.7 or 4.8), but the difference is in the value used as a benchmark, i.e. X^{nom} which is different from the average $\bar{X}$.

Please note that the measures of spread are always non-negative, and a value of zero (almost never achieved in practice in industry) should be treated as suspicious, usually indicating a lack of accuracy in acquiring the observations/measurements.

4.1.1.3 *Reliability of samples*

Taking a sample and observing it will generate a dataset. However, the way the sample is taken or chosen may have a detrimental effect on the reliability of the dataset in terms of a lack of representativeness, bias, etc. For example, in the inspection of a pallet of bricks, if we only take a few items from the top of the pallet, they might be good, but at the bottom, they might all be broken/cracked.

Another common mistake is when the dataset is generated from mixed samples. For example, consider a study on the weight of students. If we take a random sample of all students, we would expect a normal distribution of the data. However, on plotting the data, we may observe a different picture, as shown in Figure 4.9. Obviously, the dataset is made of two groups of data (samples) having different sample characteristics, such as

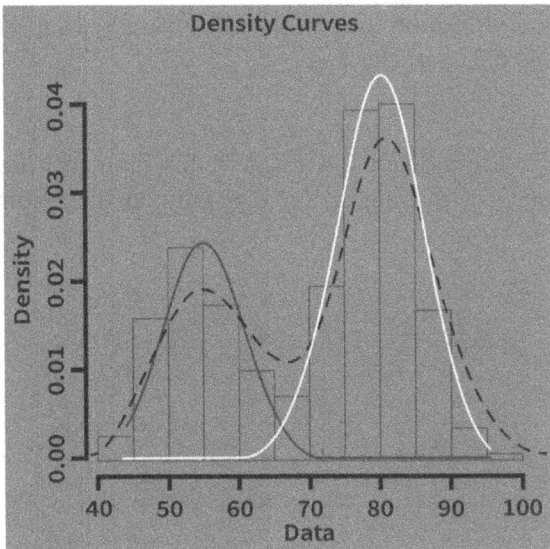

Figure 4.9.   Mixture of two samples.

sample size, average and SD. After analysing the data, we may realise that the main reason for this abnormality is the gender difference, i.e. the average male student weighs roughly 80 kg, while the average female weighs about 55 kg. In cases like this, we may consider *data stratification*. 'Stratify' means to classify or separate a population into groups graded according to a status as variously determined by location, nature, type, etc.

4.1.2 *Why is working with samples better than individual observations/measurements?*

Activity 4.1:

Consider the following scenario:

A process is being monitored by inspecting one part hourly. Initially, the process average is $\mu = 0$ and the standard deviation is $\sigma = 2$, as shown in Figure 4.10. As long as the inspector finds the measurement within the specification limits of ±6, the production is allowed to continue. If a measurement appears outside these limits, the production is halted and corrective action is taken.

Suppose a hidden, sudden change in the process causes the average to shift upward by a distance of six units (a 3σ change), causing half the process output in the current batch to now be outside the limit, as shown in Figure 4.11.

What is the probability that an inspector would catch this enormous change with the next part checked?

Answer:

Because half of the produced items are still within specification, there is a 50% chance of picking up for checking an item from inside the

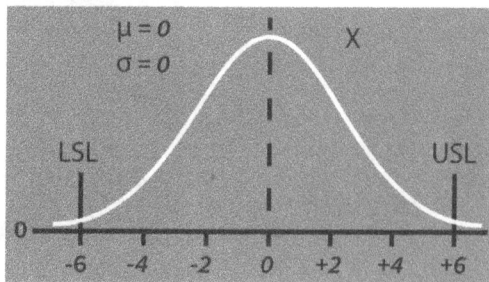

Figure 4.10.   Original process centred at zero.

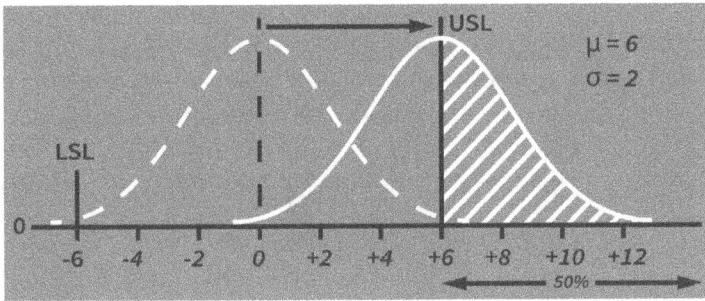

Figure 4.11. Process output after an upward shift of six units in respect to the process average.

tolerance and mistakenly deciding to continue running this modified process.

Using a sample of two ($n = 2$) items instead of one would reduce the chance of a mistake of both selected items being from the good part of the production to 50% × 50% = 25%. A sample of three items ($n = 3$) would bring down the chance to 12.5%. A sample of five ($n = 5$) would change it to only about 3%, and a sample of seven ($n = 7$) would minimise the chance of a mistake to below 1%.

These simple calculations clearly demonstrate the power of sampling.

4.2 Deviations and Errors

As explained in Chapter 3, process variations are inevitable. These variations cause deviations in the quality parameters of the products, which could be considered as *errors*. Usually, the errors are classified into two main groups.

4.2.1 *Random errors*

A *random error* is a component of the total error, which, in the course of a number of measurements, varies in an unpredictable way. Random errors can occur for a variety of reasons all acting together, such as ambient conditions (vibrations, humidity, temperature, etc.), mains power variation and lack of equipment accuracy (also discussed in Chapter 2).

Random errors can be estimated and sometimes reduced but they cannot be compensated/eliminated. Random errors are normally easily detectable as data variation, which can be measured by any measure of spread. For example, a sharp shooter aiming and repeatedly shooting at a long-distance target (Figure 4.12) will always end up with some variation (grouping) due to random errors influenced by wind, rain, weapon instability, etc.

4.2.2 *Systematic errors*

Systematic errors tend to shift all measurements in a systematic way so that over the course of a number of measurements, the mean value is constantly displaced or varies in a predictable way. The causes may be known or unknown but should always be corrected for when present. For instance, no instrument can ever be calibrated perfectly, so when a group of measurements systematically differ from the value of a standard reference specimen, an adjustment in the values should be made. The systematic errors can be detected only if there is a reference to a target. For example, if a sharp shooter systematically hitting to the top/left quadrant cannot see the point of impact in respect to the target, he would not know how to correct his aiming. So, it is important to understand that the systematic errors can be corrected for only when the 'true value' (such as the value assigned to a calibration or reference specimen) is known.

4.2.3 *Outlying errors*

During inspection, there are frequent occasions when the data taken from a study of the process population includes observation values (too small or

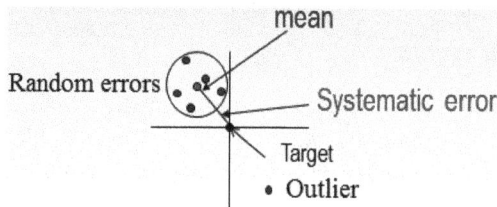

Figure 4.12.   Random, systematic and outlying errors.

too big) that appear to be totally out of context with the rest of the data. Often, these errors are due to typos, false measurement, human negligence, etc. but could also be due to the production process itself. Such blunders, usually referred to as *outlying errors* (*outliers*), significantly deviate from the rest of the data and if left undetected would compromise the descriptive statistics of the data. When the anomaly is a single measurement and is well out of range of the other data, two courses of action are possible:

(1) If the reason for the aberrant data can be traced (e.g. data entry, measurement error, poor handling i.e. unusual properties or conditions), you can legitimately replace (re-measure) or remove the outlier.
(2) If the reason cannot be identified, you should not remove it, but document this fact.

Usually, outliers are easily spotted when they are totally out of context of the rest of the data. However, in some cases, for one reason or another, this may not be so obvious. A curtain criterion should be used when it is not easy to distinguish if a given value is definitely out of the range of the other data. According to the 'Three standard deviations' (3S) criterion, the value X_i is an outlier if it is more than 3S away from the mean.

Analytically, it is expressed by the following simple formula:

$$\left| X_i - \bar{X} \right| > 3S. \tag{4.11}$$

Note the straight brackets in the above formula indicating absolute value of the difference from the mean.

From the definition of outlier, logically, when testing a dataset for outliers, you do not have to check every single value but only the extremes (usual suspects) X_{min} and X_{max}, and if they are found not to be outliers, then it means that there are no outliers at all.

Activity 4.2:
A sample of 100 parts has been inspected for a given quality characteristic, which is found to have an average $\bar{X} = 10$ and a standard deviation $S = 1$. Check if $X_{min} = 6$ and $X_{max} = 13$ are outlying errors.

This criterion seems to be very simple and easy to apply, but unfortunately, it does not work well with small samples. For example, try testing for outliers the following dataset (C):

$$C\ (10, 9, 9, 10, 10, \mathbf{30}, 11, 8, 9, 10, 9).$$

Obviously, one of the values (30) seems to be far away from all the others and so should be rejected as an outlier. However, the $3S$ criterion (based on sample SD formula) fails to detect it.

An alternative, more powerful tool for detecting outliers is the *Chauvenet's criterion*. According to this, values can be considered for rejection if the probability, P, of obtaining their deviation from the mean is less than $1/(2n)$, where n is the sample size, see Figure 4.13.

Analytically, it means that if a suspected value X_i satisfies the following condition, it should be considered for rejection as an outlier:

$$\frac{\left|X_i - \bar{X}\right|}{\sigma} > \left(d_{max}\,/\,\sigma\right), \qquad (4.12)$$

where d_{max}/σ (often denoted by t) is the ratio of the maximum acceptable deviation to standard deviation and is tabulated in relation to the sample size n. Often, d_{max}/σ is denoted by t, which allows (4.12) to be transformed to

$$\left|X_i - \bar{X}\right| > tS. \qquad (4.13)$$

This is why this criterion is also known as *tS criterion*. As you can see from Table 4.1, for large samples, the value of t is close to 3, with such cases making tS and $3S$ criteria almost identical.

To find the values of t in Excel, use the function

ABS(NORM.S.INV(1/(4n))).

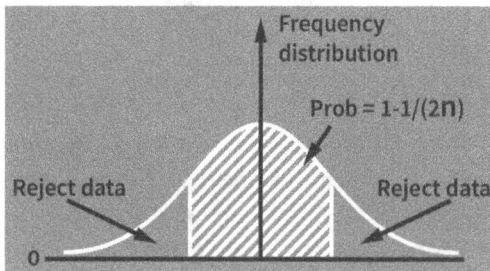

Figure 4.13.   Chauvenet's criterion for detecting outliers.

Table 4.1. Ratio of the maximum acceptable deviation to standard deviation in relation to the sample size n.

Sample Size n	Ratio d_{max}/σ, or t
3	1.38
4	1.54
5	1.65
6	1.73
7	1.80
10	1.96
15	2.13
25	2.33
50	2.57
100	2.81
300	3.14
500	3.29
1000	3.48

Activity 4.3:
Test dataset C for outliers using tS criterion. Compare with the result from $3S$ criterion.

Activity 4.4:
Compare the mean against the median as measures of location in respect to how badly they might be affected by the presence of eventual outlying errors in the dataset.

4.3 Testing of Distribution Hypothesis

Most of the formulas and conclusions in quality management are based on the assumption that the data come from a population with a specified probability distribution. This must be proved, otherwise the conclusions made could be false. In order to do the probability calculations correctly, we need to know the type of distribution. In most of the cases (including Chapter 4), we assume a normal distribution, but is this really always the case?

Histograms, control charts and probability plots are useful graphical techniques for studying the form of a density function. For example, in order to prove the normal distribution hypothesis (null hypothesis H_0 that the data follow a normal distribution), we can evaluate the closeness between curves 1 and 2 in Figure 4.14.

However, these graphical methods are subject to the criticism that decisions based on them are subjective. Two people studying the same graph may well reach different conclusions, and this is unsatisfactory in many applications. A formal test of a goodness-of-fit hypothesis has the merit that it is much more objective. For conducting the 'χ^2 goodness-of-fit test' (χ is often pronounced 'chi'), you would need to calculate the χ^2 value. The value of χ^2 is a statistical measure of the relative deviation between the empirical (observed, actual) frequency (data) f_{aj} and the expected (theoretical) distribution frequency f_{tj}:

$$\chi^2 = \sum_{j=1}^{m} \frac{\left(f_{aj} - f_{tj}\right)^2}{f_{tj}}, \qquad (4.14)$$

where

f_{aj} is the actual (empirical) frequency counted from each class of the histogram (represented by curve 1 in Figure 4.14) and

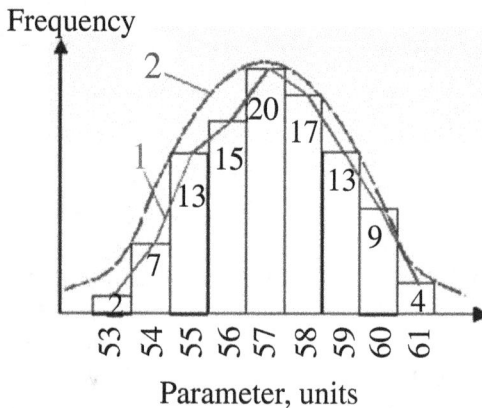

Figure 4.14.   Curve 1: empirical frequency; curve 2: theoretical frequency (plot of the probability density function).

Table 4.2. General (any type of distribution) table for calculation of χ^2.

Class	Class Intervals	Actual Frequency f_a	Theoretical Frequency f_t	$\dfrac{(f_{aj} - f_{tj})^2}{f_{tj}}$
1	(from ... to)			
2				
...				

f_{tj} is the theoretical frequency calculated from the probability density function (curve 2).

A high value (compared to the tabulated value for χ^2) would imply a poor fit between the observed and expected frequencies.

Generally, for any type of distribution, the calculations of χ^2 for any type of distribution when tested could be organised, as shown in Table 4.2.

$$\chi^2 = \sum_{j=1}^{m} \frac{\left(f_{aj} - f_{tj}\right)^2}{f_{tj}}.$$

If the calculated χ^2 value is found to be lower than a predefined benchmark value (tabulated χ^2), it is an indication of a good fit between the empirical and the normal distribution as the difference between them is statistically insignificant, so it can be concluded that the data follow a normal distribution.

$$\chi^2 = \sum_{j=1}^{m} \frac{\left(f_{aj} - f_{tj}\right)^2}{f_{tj}} < \chi^2_{\alpha,\upsilon} \Rightarrow \text{accept the hypothesis,} \qquad (4.15)$$

where

$\chi^2_{\alpha,\upsilon}$ is the tabulated χ^2 (see Table B.3 in Appendix B);

α is *significance level*, a parameter that affects the *level of confidence* (*1–α*). Usually, in the industry, a 95% level of confidence is used, so $\alpha = 5\%$ or 0.05;

υ is degree of freedom, $\upsilon = m - r - 1$, where m is the number of classes having frequency equal or bigger than five, r is the number of parameters of the distribution, for normal distribution $r = 2$ as the normal distribution has two parameters (μ, σ).

The complete procedure for running a χ^2 test is shown in Figure 4.15.

In order to avoid complicated calculations, when testing for normality by χ^2 *goodness-of-fit test'*, the data could be organised as in Table 4.3, which is tailor made for testing normal distributions.

In Table 4.3:

L_j and U_j are lower- and upper-class interval limits, respectively;

X_{mj} is the middle point (average of L_j and U_j) of the class;

Z_j is the Z transformation applied to X_{mj};

S is the estimated/sample standard deviation;

N is sample size (all items but excluding the outliers if any);

d is width of the class interval, $d = U_j - L_j$;

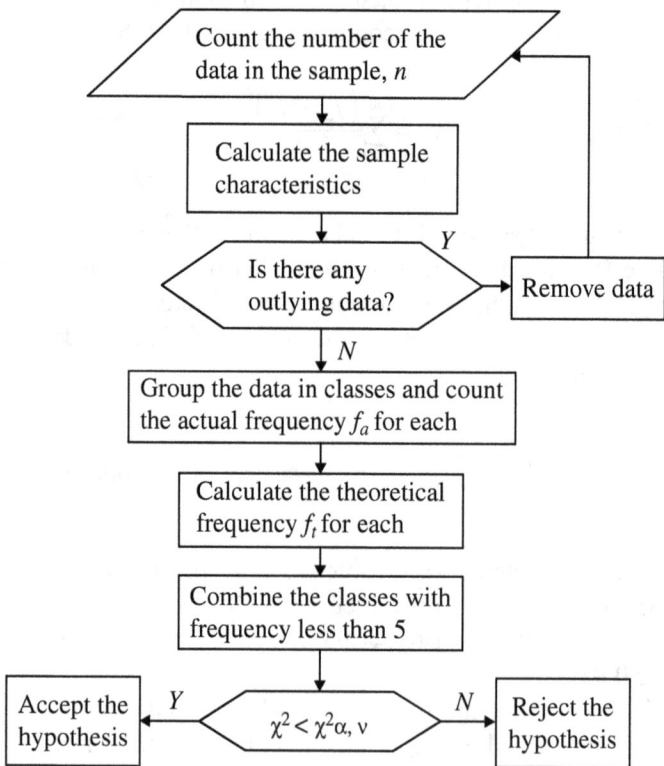

Figure 4.15. Procedure for χ^2 goodness-of-fit test.

$f(z)$ is the standard normal probability density function value for z. The values of the standard normal probability density function are tabulated (see Table B.4 in Appendix B) but could also be acquired in Excel using the function NORM.S.DIST(z, FALSE);

f_{aj} is the actual frequency;

f_{tj} is the theoretical frequency.

Note 1: If properly calculated, the f_{aj} and f_{tj} columns should have the values of their sums (at the bottom of the table) fairly close. Please check!

Note 2: χ^2 goodness-of-fit test requires frequencies in all classes to be at least five. If any class frequency (f_{aj} or f_{tj}) is smaller than five, both frequencies have to be combined with the frequency of the neighbouring class. For example, see the greyed areas in Table 4.3.

f_{aj}' and f_{tj}' are the frequencies after combining the classes with a frequency less than five to the next class. Please note that both frequencies are combined no matter which one is below five. If that combination is not enough to get both frequencies to five or above, then the combination continues with the next neighbouring class. If any class combination is needed/done in the above table, then f_{aj}' and f_{tj}' should be used in the formula in the header of the last column instead of the original frequencies f_{aj}, f_{tj}.

Table 4.3. χ^2 calculation for normal distributions.

Class No. 1–j	Class Interval $L_j \le X_j < U_j$	Middle X_{mj}	$z_j = \dfrac{Xmj - \overline{X}}{S}$	$f(z_j)$	$f_{tj} = f(z_j)\dfrac{d}{S}.N$	f_{aj}	f_{tj}'	f_{aj}'	$\dfrac{(f_{aj} - f_{tj})^2}{f_{tj}}$
1									
2									
3									
...									
j–1									
j									
					Σf_{tj}	Σf_{aj}			$\Sigma = \chi^2$

Activity 4.5:
Test if the following sample data are normally distributed (Table 4.4). Follow the procedure shown in Figure 4.15 and use Table 4.3 for the calculations. First, check for outliers. A comprehensive answer will include graphical support in terms of a histogram with theoretical and empirical curves on it.

Table 4.4.   Sample data set

18.0	16.2	15.3	12.0	12.5	15.5	13.8	17.6	14.8	15.8
18.7	17.6	14.2	15.8	13.4	17.0	14.6	17.2	17.4	17.0
19.2	15.2	16.0	19.4	14.0	17.5	16.3	17.0	22.5	15.4
16.8	14.3	14.6	130.0	16.4	11.5	14.8	12.9	16.5	18.2
15.5	18.2	15.9	18.4	13.5	15.4	12.5	16.3	18.2	14.8
16.1	15.5	17.3	16.6	18.1	14.6	14.9	18.9	12.8	15.1

Chapter 5

Process Capability

5.1 What is Process Capability?

When a process performance is measured, using Statistical Process Control (SPC) techniques, under tightly controlled conditions, the results may be used to establish a baseline for comparison. A popular baseline is the so-called 'process capability' (PC). In other words, PC is a benchmark for its performance.

So, PC is a comparison between the output of an in-control process to the specification limits (tolerance). The comparison is made by forming the ratio of the spread between the processes upper and lower specification limits (the specification 'width' equals USL–LSL) to the spread of the process values (process variation V), as measured by six process standard deviation units (the process 'width'). In other words, PC is the link between design and production because it measures how capable the production process is in meeting the design tolerance. There are two constraints in this measurement: the *design tolerance* (limits) that are set in accordance with the specification (by means of a critical parameter) and the *process variation* of the machine or equipment that will be used to make the item concerned.

The outputs of any process will vary, as discussed earlier, and it is common for specification limits to be defined so that if the measured output of the process exceeds the specified limits, the process is deemed to have failed.

We say that a process is capable if the values of (all) the important process parameter(s) falls(fall) virtually within its(their) specification limits.

5.2 Assessing Process Capability

Visually, we can assess PC by plotting the process specification limits on a histogram of the observations. If the histogram falls within the specification limits, then the process is capable. This is illustrated in Figure 5.1.

If the histogram falls beyond both specification limits (Figure 5.2), then the capability is unsatisfactory as the process produces some scrap, represented by the grey bars.

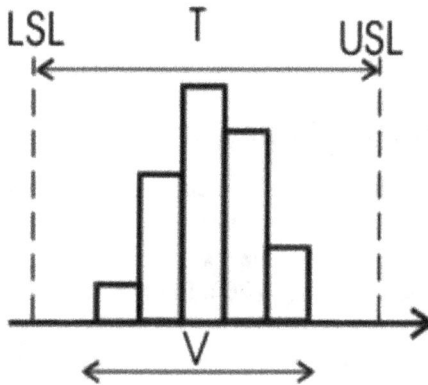

Figure 5.1.   Capable process.

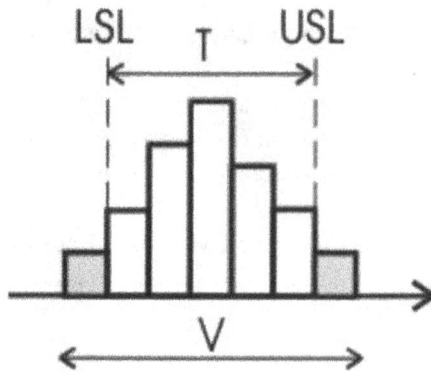

Figure 5.2.   Unsatisfactory process.

5.2.1 *Process capability index*

Numerically, we measure capability with a capability index. The general equation for the capability index, C_p, is

$$C_p = \frac{T}{V} = \frac{USL - LSL}{6\sigma}. \tag{5.1}$$

The recommended minimum value for C_p is 1.33.

If $C_p > 1.33$, the potential performance of the process is very good.

If $1 < C_p < 1.33$, the potential performance of the process is satisfactory.

If $C_p < 1$, the potential performance of the process is unsatisfactory and relevant measures should be undertaken.

However, the ultimate measure of the process performance is the percentage of rejects (items with important parameters falling outside of the tolerance), which opens the following question:

Is it possible for a process with a good capability to have a poor performance and thus produce many rejects?

5.2.2 *Adjusted process capability index*

Obviously, C_p is a simple and fast measure for PC. The only problem with the C_p index is that it does not account for a process that is off-centre (Figure 5.3).

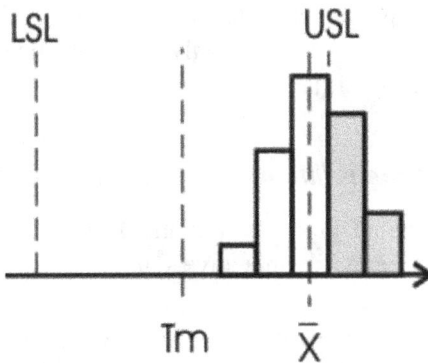

Figure 5.3. Non-centred process.

Note how the process average is shifted in respect to the target (T_m, middle of the tolerance, where T_m = (USL + LSL)/2) and although the process variation is smaller than the tolerance width, there will be a substantial amount of scrap produced. This is an example of a capable process having a poor performance. So, C_p *can only measure the process potential for a good performance.* Obviously, in order to estimate the actual process performance, a better measure is needed to account for the process setting/centring in respect to the target.

To account for off-centre processes, the ultimate *corrected/adjusted PC index C_{pk}* must be introduced, which is determined as follows:

$$C_{pk} = C_p(1 - k),\qquad(5.2)$$

where
   k is the degree of bias/correction, calculated as

$$k = \frac{|T_m - \bar{X}|}{T/2};\qquad(5.3)$$

T_m is the tolerance middle:

$$T_m = (USL + LSL)/2.\qquad(5.4)$$

The addition of k in C_{pk} quantifies the amount by which a distribution is off-centred. In other words, it accounts for shifting the average of the output in respect to the target. A perfectly centred process, where the mean is the same as the tolerance midpoint, will have a k value of zero.

A process is considered capable and well centred if $C_{pk} > 1$.

Please note that the C_{pk} index is only a reliable measure if the sample distribution is close to normal.

5.2.3 *Evaluating the relationship between C_p and C_{pk}*

Note the absolute value brackets (modulus) in the numerator in (5.3), which means that k cannot be negative. Therefore, it is always true that $C_{pk} \le C_p$.

At this point, we naturally question: what if $C_{pk} < 1$? Does it mean that there is a lack of process potential capability ($C_p < 1$), or is it an indication of poor process setting or both? This question is illustrated in Figure 5.4. How can we find out?

Figure 5.4. Car crashed in the garage. Was the garage too tight? Or was the car poorly centred?

The following activities could help to answer these questions.

Activity 5.1:
Try to guess the magnitude ($<$/$>$1) of the values of both C_p and C_{pk} for the cases shown in Figures 5.1–5.3.

Activity 5.2:
Estimate the PC if the specified tolerance is $\varnothing 30\,^{+4}_{0}$ and from a sample, it has been calculated that the mean $\bar{X} = 30$ and SD = 0.5.

Activity 5.3:
Is it possible that $C_p = C_{pk}$? Explain.

Activity 5.4:
Could C_{pk} be a negative number? Explain.

Things to remember:

- C_p is only a measure of variation.
- C_{pk} is a measure of both location and variation.
- C_p will never be a negative number.

Chapter 6

Control Charts

6.1 What Do We Need Control Charts For?

As explained in previous chapters, all processes, regardless of how well designed or carefully maintained they are, will always have a certain amount of inherent or natural variability. This natural variability is the cumulative effect of many causes acting together on key quality characteristics and changing the output of the processes. These changes can exist for two main groups of reasons:

1. Common (random) causes are flaws inherent in the nature of the process, such as power and temperature variation, vibrations and material inconsistency. They cannot be eliminated and are a function of the process capability. Variations in process will follow a normal distribution, which makes the process stable and predictable, as shown in Figure 6.1.

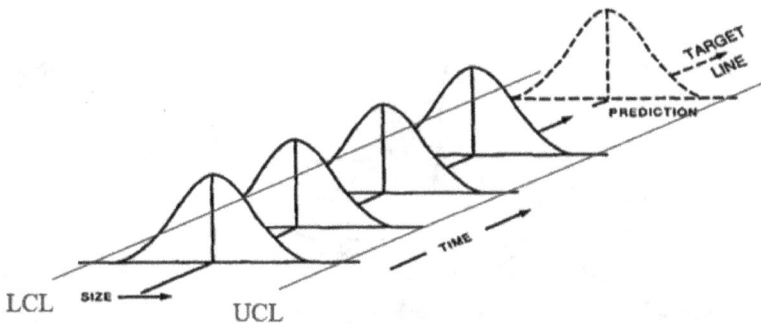

Figure 6.1.   In-control process with only common causes of variation.

The output of such a process forms a stable and predictable distribution over time, and the process is said to be *in statistical control.*

2. Special causes are variations caused by unusual circumstances or events, such as broken tools, operator errors and defective raw material. Such a variability is generally larger when compared to the usually small variation caused by common random causes, and it represents an unacceptable level of process performance. We refer to these sources of variability that are not part of the natural cause pattern as *special causes of variation.* A process that is operating in a random presence of special (often systematic) causes is said to be an *out-of-control process.* Variations due to a systematic cause inflict an out-of-control condition, making the process unstable and unpredictable (Figure 6.2).

However, as these kinds of causes are avoidable, the out-of-control conditions are rectifiable. Therefore, when a special cause is detected, an intervention is required.

In fact, the worst-case scenario is a combination of the two types of causes, where we may have a period of consistent, stable output followed by an abrupt change, which may go unnoticed.

In order to guarantee the quality of a product, variation has to be predictable and controlled within certain limits (tolerance). If you want to reduce the amount of variation in a process, you need to be able to dynamically monitor the results/changes of the process. For example, a histogram built on the basis of a relatively big sample size is a powerful tool

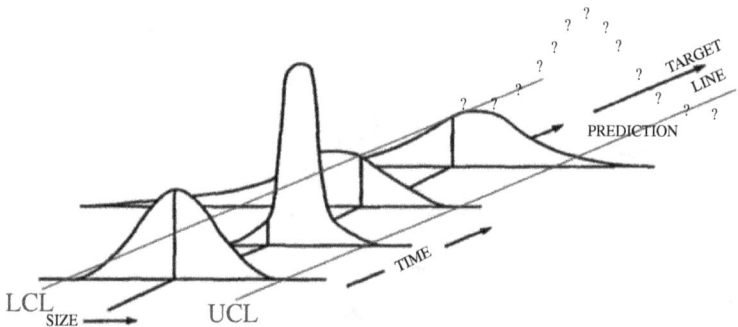

Figure 6.2.   Out-of-control process operating with special causes of variations.

for quality control (QC) but is mostly suitable for 'offline' control used for incoming goods control or outgoing products control, i.e. when production is already finished, which may be too late for corrections. But can we always build histograms to dynamically monitor a process? Obviously, we need a faster tool for online (in-process) QC to monitor an ongoing production process. Control charts (CC) are very good at this because they are much faster due to the fact that they work with small samples. CC can quickly distinguish causes of variation, which are either special or common. Therefore, the purpose of building a CC is for the quick detection of any adverse changes in the process, causing changes in the observed parameter.

6.2 What is Control Chart?

CCs were first used by Walter Shewhart in 1924.

A CC is a graph or chart of a quality characteristic measured/computed from a sample versus the sample number/time with displayed limit lines. The lines are called control lines, see Figure 6.3.

There are three kinds of control lines: upper control limit (UCL), lower control limit (LCL) and central line (CL).

When a process is in control, we expect a certain distribution of the points in the CC. In the central green 1σ zone (Figure 6.3), around 68% of all points is normally expected. In the yellow 2σ zone, around 28% and 3σ zone, 4% are expected.

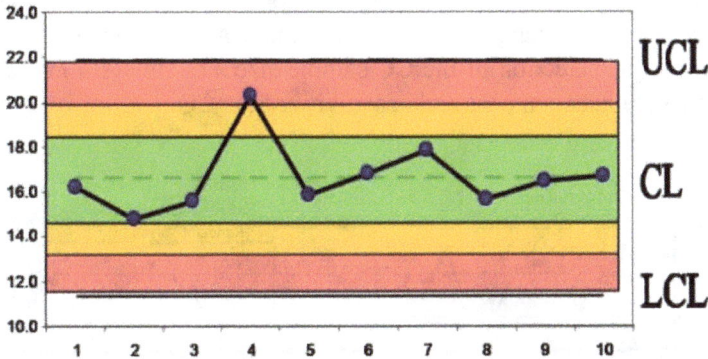

Figure 6.3.   Control chart.

6.3 How Does a Control Chart Work?

Any problems/changes in the process are signalled by abnormalities (e.g. points beyond the limits) in the graph in Figure 6.4(a). When the problem(s) is (are) fixed and the process is in control, all points are within the limits (Figure 6.4(b)) following the distribution explained above. If an improvement leading to a reduced variation (e.g. better machine, improved technology, raw materials) is introduced, the control limits of the CC must be re-calculated (see UCL' and LCL' in Figure 6.4(c)) in order to keep the sensitivity of the change detection.

So, a CC will tell us that one of the following states applies:

EITHER:
The process control has been achieved at the desired level and the process therefore should be left to continue.

OR:
The process is not achieving the desired level, so the process is 'out of control', and intervention is required.

6.4 Sampling Strategy for Control Chart

The samples needed to build a CC can be taken at a certain period of time, e.g. checking a sample every hour, which is applicable to regular running/continuous production or on lot-by-lot basis, e.g. checking a sample out of every 200 produced (intermittent production).

The sample size is another important issue in quality control. A bigger sample size will have higher discriminating power (steeper curve, Figure 6.5) to detect adverse changes, but this will incur a higher inspection cost. Obviously, the selection of the CC sample size n is always a matter of a good balance between cost and sensitivity of the CC.

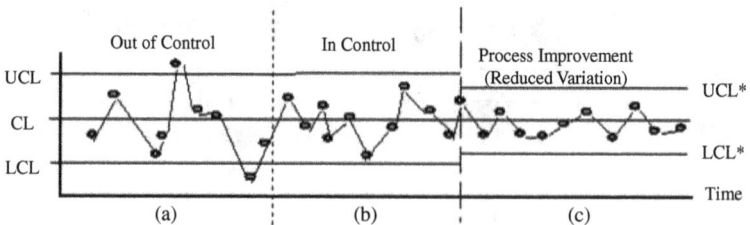

Figure 6.4.   CC with are the original CLs (left) and revised CLs (right).

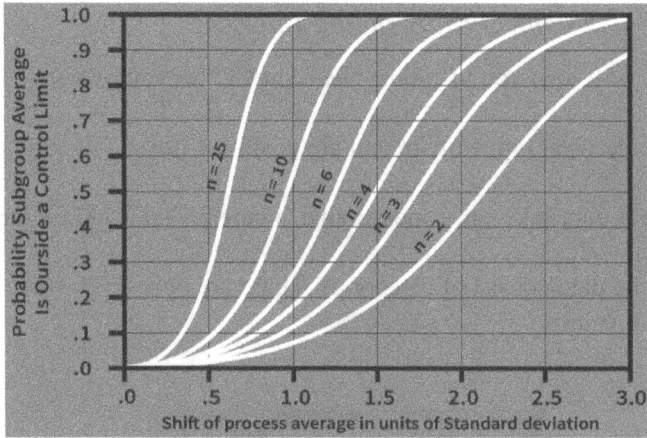

Figure 6.5.   Sensitivity of CCs to detect changes depending on the sample size *n*.

6.5 Types of Control Chart

Depending on the nature of the data, there are two main types of CCs: those for *variables* and those for *attributes*.

The group of *variable CCs* (see Figure 6.6) is used when the data are generated through measurements of quantitative performance characteristics (variables). Values are expressed through some measuring unit (mm, kg, etc.).

The group of *attribute CCs* is used when we deal with qualitative performance characteristics (attributes) or count the number of defects or defective items. Values are determined on 'good-or-bad-' and 'go-or-not-go-'type conditions from which a measure of defect proportion is made.

There are also some specialised charts for variable data. They can be helpful but are more difficult to use and explain, so they are not covered in this book. Here's a quick list:

- The cumulative sum (CuSum) chart is more sensitive to small, sustained changes in level than the standard control charts.
- The exponentially weighted moving average (EWMA) chart is also sensitive to small, sustained changes in level. It smooths out noisy datasets and is sometimes used with autocorrelated data.
- The median and individual measurements (MI) chart can control an entire family of processes in one chart. It's useful for multi-cavity moulds and multi-head fillers.

6.5.1 *Which control chart should I use?*

Table 6.1 will help you to choose the right CC for your case.

First, decide what type of data you are dealing with:

- Variable data take on a measurable, numeric value. There are many possible values.
- Attribute data consist of categories. There are only a few (usually two) discrete values.

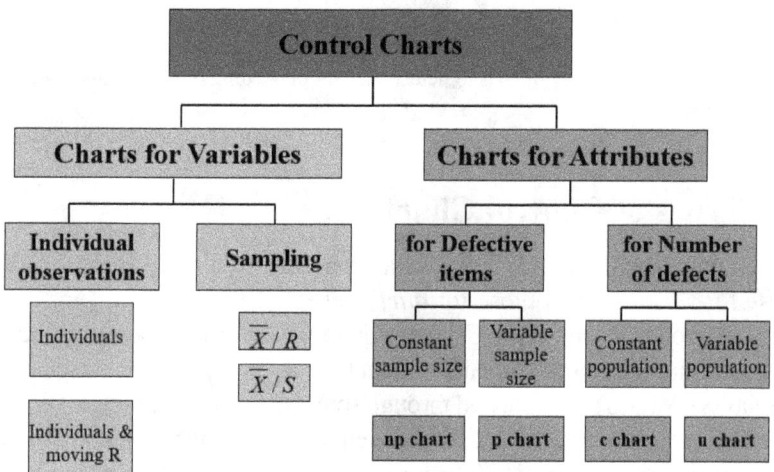

Figure 6.6.   Types of CCs.

Table 6.1.   How to choose the right CC?

Data Type	Distribution	Subgroup Size	Control Chart
Variable	Any	1	Individuals and moving R
		2–10	$\bar{X}/R$
		More than 10	$\bar{X}/S$
Attribute	Binomial	Constant	np
		Varying	p
	Poisson	Constant	c
		Varying	u

With variable data, decide how large your subgroups are:

- If the subgroup size is 1, then use an individual measurements chart with or without a moving R (range) chart.
- If the subgroup size is from 2 to 10 (or possibly 12), then use an $\bar{X}$-and-R-chart.
- If the subgroup size is over 10 (or 12), then use an $\bar{X}$-and-S-chart.

With attribute data, decide on what type of distribution the data follows:

- Binomial data take on two values, usually 'good' or 'bad'. If the sample size is constant, use an np-chart. If the sample size changes, use a p-chart.
- Poisson data are a count of infrequent events, usually defects. If the sample size is constant, use a c-chart. If the sample size changes, use a u-chart.

6.5.2 *Control charts for variables versus control charts for attributes*

The main advantage of variable control charts is their sensitivity to detect adverse changes. For this reason, they are also known as 'leading indicators'—they may alert us before the number of rejects (scrap) increases in the production process. However, as most of them work in pairs, they are slightly harder to interpret.

The main advantages of attribute control charts are that they are easier and cheaper to do and to acquire data as most times, it is just counting. There is no need for measuring tools. In addition, they are easier to interpret for people unfamiliar with statistics and QC. However, usually, the attribute CCs require a bigger sample size, typically above 50.

6.6 Control Charts for Variables

Most popular CCs for variables are a plot of usually two parameters derived from the samples statistics on a pair of scales, typically located above and below each other, running horizontally, see Figure 6.7. These statistics are specially paired so that one measures the location and the other measures the spread of the data in the sample, e.g.:

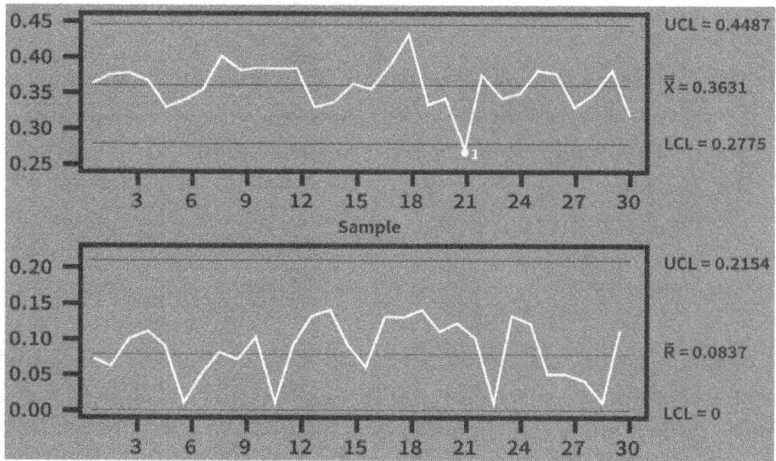

Figure 6.7.   Example for variable CC.

- sample mean/sample range ($\bar{X}/R$),
- sample mean/sample SD ($\bar{X}/S$),
- sample mean/variance ($\bar{X}/S^2$),
- sample median/sample range ($\tilde{X}/R$).

6.6.1 *How to make control charts for variables?*

The following steps describe the procedure for constructing a variable CC:

(1) Select the type of the control chart, e.g. $\bar{X}/R$, $\bar{X}/S$.
(2) Choose a sampling strategy and sample size , e.g. take a sample of five every hour, on the hour.
(3) Establish the control lines; calculate control lines: CL, UCL and LCL. Formulae for calculations are given in Table 6.2.

In Table 6.2,

$\bar{\bar{X}}$ (double bar) is the average of the sample averages ($\bar{X}$),

$\bar{R}$ (R bar) is the average of the sample ranges (R),

N is the number of consecutive samples taken over time and coefficients A_2, D_3 and D_4 depend on the sample size n, as shown in Table 6.3.

Table 6.2. Control limits formulae.

$\bar{X}$ Control Chart	R Control Chart	S Control Chart
$CL = \bar{\bar{X}} = \dfrac{\sum \bar{X}_i}{N}$	$CL = \bar{R} = \dfrac{\sum R_i}{N}$	$CL = \bar{S} = \dfrac{\sum S_i}{N}$
$UCL = \bar{\bar{X}} + A_2 \bar{R}$	$UCL = D_4 \cdot \bar{R}$	$UCL = B_4 \bar{S}$
$LCL = \bar{\bar{X}} - A_2 \bar{R}$	$LCL = D_3 \cdot \bar{R}$	$UCL = B_3 \bar{S}$

Table 6.3. Control limits coefficients.

n	A_2	D_4	D_3	B_3	B_4
2	1.880	3.267	0	0	3.267
3	1.023	2.575	0	0	2.568
4	0.729	2.282	0	0	2.266
5	0.577	2.114	0	0	2.089
6	0.483	2.004	0	0.030	1.970
7	0.419	1.924	0.076	0.118	1.882
8	0.373	1.864	0.136	0.185	1.815
9	0.337	1.816	0.184	0.239	1.761
10	0.308	1.777	0.223	0.284	1.716

Please note the following:
(a) Ideally, the CC limits must be estimated in advance before you start plotting any points from preliminary samples taken when the process is thought to be in control. These estimates should usually be based on at least 20 samples when the process is deemed to be running in control. Once these limits have been established, they are supposed to be kept until there is a change (e.g. new machine, new type of workpiece) in the process.

(b) Obviously, in the classic form of CC for variables involving samples, the calculation of the control limits is not analytically related to the design tolerance, which might cause a poor initial centring of the process to pass undetected resulting in a substantial number of rejects. In order to link the $\bar{X}$-CC control limits to the tolerance limits, it is often suggested [6] that the $\bar{X}$ in the formula in Table 6.2 is replaced by the middle of the tolerance T_m.

(4) Construct/plot the CC limits using appropriate scaling factors in order to facilitate the interpretation.
(5) Start taking samples and recording data, plotting, e.g. $\bar{X}$ and R values as computed for each sample.
(6) Alongside, add any relevant important information, e.g. the nature of data, period/time when samples were taken, instruments used, person's responsibility.

Activity 6.1:
Using the data from Table 6.4, a $\bar{X}/R$-CC has been plotted in Figure 6.8. Check if the CC has been calculated and constructed correctly.

Table 6.4.   Sample data set.

Part	Sample Number									
Number	1	2	3	4	5	6	7	8	9	10
1	18	16	15	17	18	12	15	14	15	17
2	20	15	15	16	14	22	18	17	15	17
3	15	11	14	13	16	17	24	15	23	16
4	16	15	18	40	17	17	17	10	15	16
5	12	17	17	17	16	16	16	22	15	18
$\bar{X}$	16.2	14.8	15.6	20.3	15.9	16.8	17.9	15.7	16.5	16.7
R	8	6	4	27	4	10	10	12	8	2

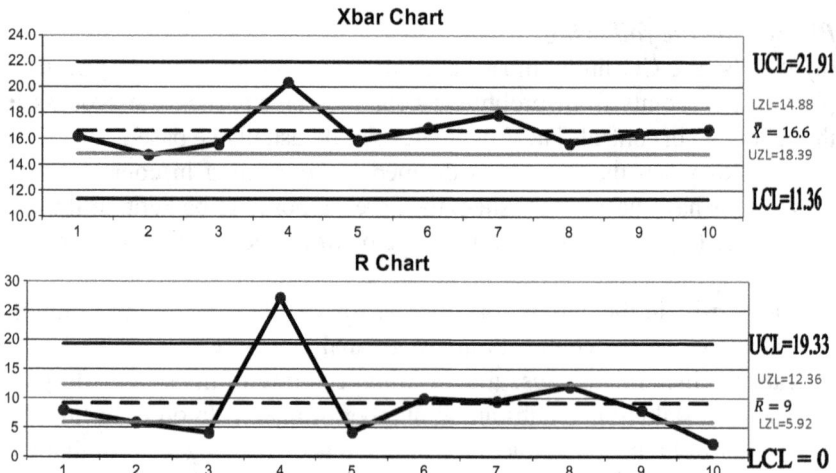

Figure 6.8. $\bar{X}/R$-CC.

6.6.2 *Control chart interpretation*

While constructing CCs is an easy routine job, interpreting them is not always that easy as it requires further knowledge about the so-called out-of-control conditions. A control chart may indicate an *out-of-control condition* when one or more points fall beyond the control limits or when the plotted points exhibit some *non-random pattern of behaviour.*

Some typical out-of-control conditions/patterns are:

(1) points outside control limits (freaks);
(2) less than two-thirds of all points in the middle one-third of control limits zone;
(3) seven consecutive points all above or below CL;
(4) eight consecutive points upwards or downwards.

Interpretation procedure:

- If any of the above four tests fail, the process is out of control for range. Investigate for special causes and correct.
- If all four tests pass, the process is in control for range. Interpret for control on mean chart.
- Only if all four tests pass in the mean chart as well, then the process is said to be in control.

Table 6.5 presents some popular abnormalities frequently observed in variable CC and their possible causes.

Activity 6.2:

(a) Interpret the abnormality given in Activity 6.1. Suggest a possible cause.
(b) Which of CC $\bar{X}/R$ or $\bar{X}/S$ is more sensitive to detect outliers?
(c) When dealing with CCs for variables, what are the relationships between control limits and specification limits?
(d) Is it possible a process which seems to be in control as monitored by CC for variables to produce a significant amount of scrap?

Table 6.5.　Popular abnormalities and their possible causes in variable CCs.

Observed Abnormality	Possible Cause (Variable CC)
A single point out of limits in R-chart	Outlying data
Many points scattered out of limits in R- or S-chart, or points out of limits in $\bar{X}$-chart	(1) Mixed data (non-homogeneous data) (2) Unstable process
A trend (up or down) in $\bar{X}$-chart	Influence of some systematic factor (in manufacturing, typically a tool wear)
A sudden 'jump' in $\bar{X}$-chart	(1) Sudden change in process setup (2) Unstable process
Seven consecutive points all above or below CL in $\bar{X}$-chart	Wrong process setting
A run of points either hugging CL (in 1σ zone) in $\bar{X}$-chart or located in the bottom part of the 3σ zone in R- or S-chart	Hidden process improvement(s)

6.7　Control Charts for Attributes

Many quality characteristics cannot be conveniently represented numerically. In such cases, each item inspected is classified as either conforming or non-conforming to the specifications of that quality characteristic. Quality characteristics of this type are called attributes. Examples are non-functional semiconductor chips, warped connecting rods and number of defects on a piece of furniture.

6.7.1　*When to use control charts for attributes?*

(a)　Control of non-quantifiable parameters;
(b)　When measurements are too difficult to take:
  •　measurements are too costly;
  •　GO, NOT-GO control;
  •　measurements interpreted as pass–fail.

Note: When working with CC for attributes, a bigger sample size (usually $n > 100$) is required.

　　The most inexpensive statistic is the yield of the production line. Yield is related to the ratio of defective vs. non-defective, conforming vs. non-conforming or functional vs. non-functional.

We often measure:

- number of defective (non-conforming) items (*pn*-chart, also known as *np*-chart);
- fraction defective items (*p*-chart);
- number of defects (non-conformities) in product (*c*-chart);
- number of defects per unit area (*u*-chart).

With attribute data, in order to decide what type of distribution the data follow, consider that:

- binomial data take on two values, usually 'good' or 'bad'. If the sample size is constant, use an *np*-chart. If the sample size changes, use a *p*-chart.
- Poisson data are a count of infrequent events, usually defects. If the sample size is constant, use a *c*-chart. If the sample size changes, use a *u*-chart.

6.7.2 *How to make control charts for attributes?*

The structure of a CC for attributes is similar to those for variables but slightly simpler given that they are not paired.

6.7.2.1 *p-chart*

p-charts are normally used when the sample size *n* is variable, so we monitor the fraction (percentage) of defective items.

Follow the steps below to make a *p*-chart:

Step 1. Collect the data, which tell the number of inspected *n* and the number of defective products *pn*. Divide the data into subgroups. Usually, the data are grouped by date or lots. The subgroup size *n* should be over 50.

Step 2. Compute the fraction of defective for each group and enter it on a data sheet. You can use a data sheet similar to Table 6.6. To find the fraction defective, use the following formula:

$$p = \frac{number\ of\ defectives}{size\ of\ sub\ group} = \frac{pn}{n}. \tag{6.1}$$

To indicate this value as a percentage, multiply by 100.

Table 6.6.   Fraction defective for electric components.

Sub-group No.	Sub-group Size, n	No. of Defectives, pn	Percent Defective, p %	UCL %	LCL %
1	115	15	13.0	18.8	1.8
2	220	18	8.2	16.4	4.1
3	210	23	11.0	16.6	4.0
4	220	22	10.0	16.4	4.1
5	220	18	8.2	16.4	4.1
6	255	15	5.9	16.0	4.6
7	440	44	10.0	14.6	5.9
8	365	47	12.9	15.1	5.5
9	255	13	5.1	16.0	4.6
10	300	33	11.0	15.6	5.0
11	280	42	15.0	15.7	4.8
12	330	46	13.9	15.3	5.3
13	320	38	11.9	15.4	5.2
14	225	29	12.9	16.4	4.2
15	290	26	9.0	15.6	4.9
16	170	17	10.0	17.3	3.3
17	65	5	7.7	21.6	−1.0
18	100	7	7.0	19.4	1.2
19	135	14	10.4	18.1	2.4
20	280	36	12.9	15.7	4.8
21	250	25	10.0	16.1	4.5
22	220	24	10.9	16.4	4.1
23	220	20	9.1	16.4	4.1
24	220	15	6.8	16.4	4.1
25	220	18	8.2	16.4	4.1
Total	5925	610			

Step 3. Find the CL as average fraction defective:

$$CL = \bar{p} = \frac{Total\ defectives}{Total\ inspected} = \frac{\sum pn}{\sum n} = \frac{610}{5925} = 0.103 = 10.3\%. \quad (6.2)$$

Step 4. Compute the control limits:

$$UCL = \bar{p} + \frac{3}{\sqrt{n}}\sqrt{\bar{p}(1-\bar{p})} = 0.103 + \frac{3}{\sqrt{n}}0.304, \quad (6.3)$$

$$LCL = \bar{p} - \frac{3}{\sqrt{n}}\sqrt{\bar{p}(1-\bar{p})} = 0.103 - \frac{3}{\sqrt{n}}0.304. \quad (6.4)$$

Step 5. Draw the control lines and plot p.

Note: CL is a constant represented by a straight line (see Figure 6.9), but since the sample size n varies, UCL and LCL are variables too.

6.7.2.2 *pn-chart*

pn-charts are normally used when the sample size n is constant.

The procedure for making *pn*-charts is similar to the one described above. Table 6.7 represents plating defects of assembled parts.

Central line (average number of defectives per sample):

$$CL = \bar{p}n = \frac{129}{30} = 4.30. \quad (6.5)$$

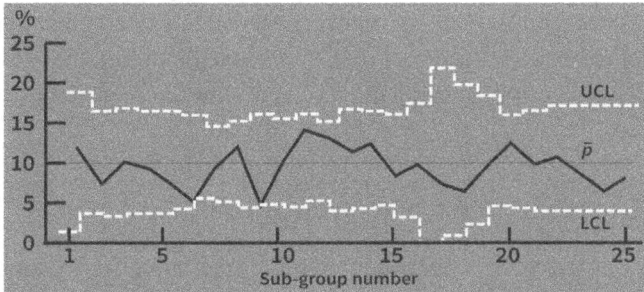

Figure 6.9. *p*-chart.

Table 6.7. Plating defects of assembled parts.

Sub-group No.	Sub-group Size, n	No. of Defectives, pn	Sub-group No.	Sub-group Size, n	No. of Defectives, pn
1	100	1	16	100	5
2	100	6	17	100	4
3	100	5	18	100	1
4	100	5	19	100	6
5	100	4	20	100	15
6	100	3	21	100	12
7	100	2	22	100	6
8	100	2	23	100	3
9	100	4	24	100	4
10	100	6	25	100	3
11	100	2	26	100	3
12	100	1	27	100	2
13	100	3	28	100	5
14	100	1	29	100	7
15	100	4	30	100	4
$\bar{p}$ = 129/3000 = 0.043			Total	3000	129
			Average $\bar{n}$ =	100	$\bar{p}n$ = 4.3

Control limits:

$$UCL = \bar{p}n + 3\sqrt{\bar{p}n(1-\bar{p})} = 4.30 + 6.09 = 10.39, \qquad (6.6)$$

$$LCL = \bar{p} - +3\sqrt{\bar{p}n(1-\bar{p})} = 4.30 - 6.09 = -1.79. \qquad (6.7)$$

Due to the nature of the data, where the LCL value is calculated and appears to be negative, the zero line is accepted as the LCL. This might be a bit confusing for interpretation because the CL is no longer central, see Figure 6.10.

6.7.2.3 *u-chart*

u-charts or *c*-charts are used when counting the number of (same or different type) defects on product.

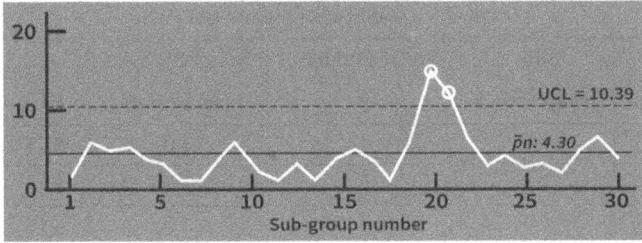

Figure 6.10.   *pn*-chart.

If the item/material being inspected is not constant in volume, area, length, etc. (such as the unevenness of woven materials or pinholes in enamel wire), a *u*-chart is normally used.

Step 1. Collect as much data as possible for number of units *n* and number of defects *c*. For example, assume there is a 5 m² electroplated copper plate with eight pin holes (defects) in it. One unit could be 1 m², so $n = 5$, and $c = 8$.

Step 2. Group the data. Do this by lots, products or samples, etc., where n = subgroup size and c = number of defects. It is recommended to set the subgroup size so that u, which represents the number of defects per unit, will be larger than two or three.

Step 3. Find the number of defects per unit and then compute u:

$$u = \frac{Number\ of\ defects\ per\ subgroup}{Number\ of\ units\ per\ subgroup} = \frac{C}{n}, \qquad (6.8)$$

$$\bar{u} = \frac{Total\ defects\ for\ all\ subgroups}{Total\ units\ for\ all\ subgroups} = \frac{\sum c}{\sum n}. \qquad (6.9)$$

An example sample data are shown in Table 6.8, representing the number of pinholes in wire:

$$\bar{u} = \frac{75}{25.4} = 2.95.$$

Table 6.8. Number of pinholes in wire.

Sub-group No.	Sub-group Size, n	No. of Pinholes, c	No. of Pinholes per Unit, u	UCL	LCL
1	1.0	4	4.0	8.15	−2.20
2	1.0	5	5.0	8.15	−2.20
3	1.0	3	3.0	8.15	−2.20
4	1.0	3	3.0	8.15	−2.20
5	1.0	5	5.0	8.15	−2.20
6	1.3	2	1.5	7.51	−1.56
7	1.3	5	3.8	7.51	−1.56
8	1.3	3	2.3	7.51	−1.56
9	1.3	2	1.5	7.51	−1.56
10	1.3	1	0.8	7.51	−1.56
11	1.3	5	3.8	7.51	−1.56
12	1.3	2	1.5	7.51	−1.56
13	1.3	4	3.1	7.51	−1.56
14	1.3	2	1.5	7.51	−1.56
15	1.2	6	5.0	7.70	−1.75
16	1.2	4	3.3	7.70	−1.75
17	1.2	0	0.0	7.70	−1.75
18	1.7	8	4.7	6.94	−0.99
19	1.7	3	1.8	6.94	−0.99
20	1.7	8	4.7	6.94	−0.99
Total	**25.4**	**75**	$\overline{X}$ **= 2.98**		

Step 4. Compute the CL (average number of defects per sample) and the control limits:

$$CL = \overline{u} = 2.95, \tag{6.10}$$

$$UCL = \overline{u} + 3\sqrt{\frac{\overline{u}}{n}} = 2.95 + \frac{5.15}{\sqrt{n}}, \tag{6.11}$$

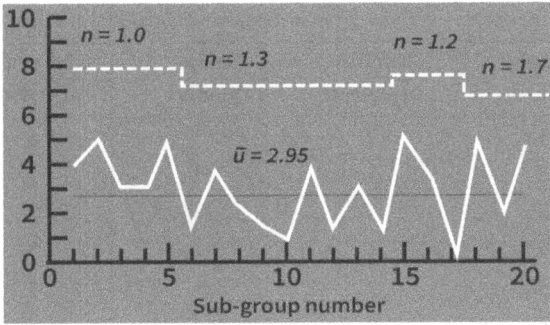

Figure 6.11.   *u*-chart.

$$LCL = \bar{u} - 3\sqrt{\frac{\bar{u}}{n}} = 2.95 - \frac{5.15}{\sqrt{n}}. \tag{6.12}$$

Note again the variable control limits and that there is no consideration to negative LCL values (see Figure 6.11).

6.7.2.4 *c-chart*

A *c*-chart is used in dealing with the number of defects which appear in fixed unit samples (identical items), such as the number of defects (cracks, stains, sinks, etc.) in, for example, bricks, tiles and cups.

Table 6.9 shows data on the number of defects in woven material (socks).

$$CL = \bar{c} = \frac{82}{20} = 4.1, \tag{6.13}$$

$$UCL = \bar{c} + 3\sqrt{\bar{c}} = 4.1 + 3\sqrt{4.1} = 10.17, \tag{6.14}$$

$$LCL = \bar{c} - 3\sqrt{\bar{c}} = 4.1 - 3\sqrt{4.1} = -1.97. \tag{6.15}$$

Again, there is no consideration due to the negative value for the LCL (see Figure 6.12).

Table 6.9.   Number of defects in woven material (socks).

Sample Number	Number of Defects	Sample Number	Number of Defects
1	7	11	6
2	5	12	3
3	3	13	2
4	4	14	7
5	3	15	2
6	8	16	4
7	2	17	7
8	3	18	4
9	4	19	2
10	3	20	3
		Total	**82**

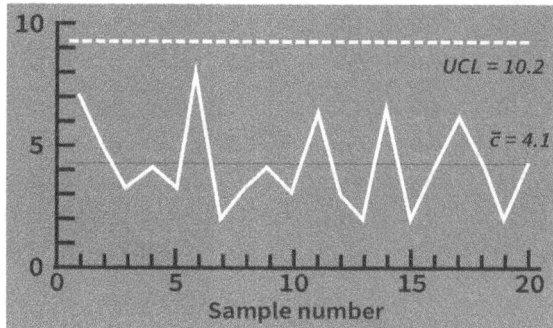

Figure 6.12. c-chart.

6.7.2.5 *Attribute control chart interpretation*

It consists of the same process as the one used in CCs for variables:

- Observe plotted points and check the rules for non-control patterns.
- Investigate for special cases and reject points if needed.
- Re-draw the CC until having an in-statistical process.

6.7.2.6 *Procedure for dealing with abnormalities*

If an abnormality appears on the chart, investigate the cause and take appropriate action. Recalculate the control lines:

- Data on points indicating abnormality, where the cause has been found and corrected, should not be included in the recalculation.
- Data on abnormal points, for which the cause cannot be found or an action is to be taken, should be included.

Activity 6.3:

(a) Interpret the abnormality given in Figure 6.10 (Table 6.7). Suggest a possible cause.
(b) Does a run of seven to eight points above or below the CL in a CC for attributes indicate the same problem (cause) as in a CC for variables?
(c) In constructing p- and u-charts, what is the relationship between the subgroup size n and the UCL–LCL range?

Chapter 7

Other Tools for Quality Management and Optimisation

7.1 Check Sheets

The check sheet is a form whose function is to compile and organise the data for future analysis. Although there is no standard layout, there are several types of popular check sheets that might be used for a variety of purposes.

7.1.1 *Production process distribution check sheet*

Usually, this type of check sheet (see Figure 7.1) is used with quantitative data (in this case, measured film thickness), and to some extent, it resembles a histogram as it records the frequency of the occurrences in each numerical category/range.

Any extra information that could be of value for the analysis could be added to the form as shown on the top of this check sheet.

7.1.2 *Defective items check sheet*

This type of check sheet (see Figure 7.2) is used with data obtained by counting (in this case, defects in moulded parts), and in some way, it resembles a tally chart as it records the frequency of the occurrences in each category.

93

Frequency distribution for film c

Data Collector:	Percival D. Oviatt	Facility:	Kodak Park Bldg 29
Date:	1/17 - 1/23	Room:	13B
Shift:	C	Apparatus:	Film coater
Product:	Ektachrome film	Process:	Deposition
Batch:	23		
Inspections:	160		

Film thickness (μm) — LSL at 93.0, USL at 105.0

Frequency distribution (X = one inspection), film thickness (μm) from 89.0 to 111.0:

Thickness (μm)	Frequency
89.0	1
90.0	2
91.0	1
92.0	2
93.0	1
94.0	3
95.0	2
96.0	5
97.0	6
98.0	9
99.0	8
100.0	10
101.0	14
102.0	20
103.0	25
104.0	14
105.0	9
106.0	13
107.0	8
108.0	4
109.0	2
110.0	1
111.0	0

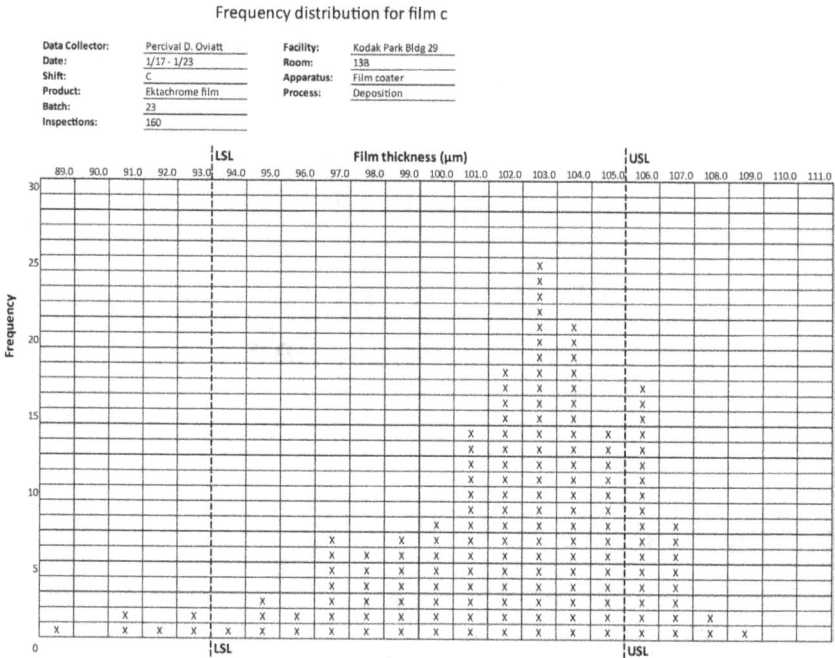

Figure 7.1. Production process distribution check sheet.

Again, any useful information could be added to the form as shown on the top of this check sheet.

In order to reduce the number of defectives, it is necessary to know what sort of defects occur and their percentages. Since different defects have different causes, it is of no use to just list the total number of defects. The defects and their frequencies must be recorded, and appropriate action should be taken to eliminate the defects, perhaps starting with that having the highest frequency.

7.1.3 *Defective location check sheet*

In some cases, the analysis would benefit by not only the nature and frequency of the defect but also the location where it would appear.

The defective location check sheet shown in Figure 7.3 depicts the location and grouping of three main types of defects identified in a painting operation of ten car doors.

Check Sheet

Product

Manufacturing stage: Final insp.

Type of defect: scar, incomplete, misshapen

Total no. inspected: 2530

Remarks: all items inspected

Date:
Factory:
Section:
Inspector's name:
Lot no.:
Order no.:

Type	Check	Subtotal
Surface scars	𝍸 𝍸 𝍸 𝍸 𝍸 𝍸 ||	32
Cracks	𝍸 𝍸 𝍸 𝍸 |||	23
Incomplete	𝍸 𝍸 𝍸 𝍸 𝍸 𝍸 𝍸 𝍸 𝍸 |||	48
Misshapen	||||	4
Others	𝍸 |||	8
	Grand total:	115
Total rejects	𝍸 𝍸 𝍸 𝍸 𝍸 𝍸 𝍸 𝍸 𝍸 𝍸 𝍸 𝍸 𝍸 𝍸 𝍸 𝍸 |	86

Figure 7.2. Defective items check sheet used in moulded parts.

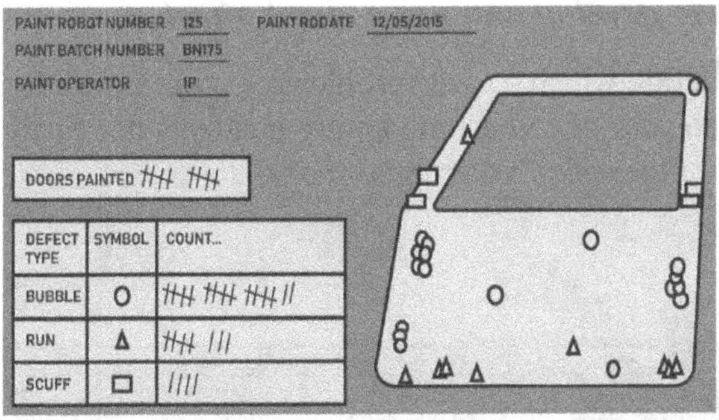

Figure 7.3. Defective location check sheet used in car door painting operation.

7.1.4 *Defective cause check sheet*

Often alongside with the frequency of the defects, it is important to analyse the cause of the defects. A useful form to organise this sort of data is the defective cause check sheet. This check sheet helps to proceed a step further in detecting the causes for these defects by making use of stratification.

Figure 7.4 shows an example of a defective cause check sheet used for recording defect occurrences in Bakelite knobs with regard to machines, workers days and types of defects.

At a glance, we can see that worker B has produced a lot of defects. On Wednesday, all workers produced many defects. Upon investigation, it was revealed that worker B has not changed dies often enough and on Wednesday, a raw material composition, which was more likely to cause defects, had been used.

7.1.5 Check-up confirmation check sheet (checklist)

This type of check sheet (also known as checklist) is used to keep a record of operations performed to make sure that nothing is overlooked. It should be applied when the list of operations is long and/or complex. All the tests/operations/items are listed beforehand on the check sheet, and a mark is made for each item as you proceed with the check-up.

Figure 7.5 shows a check sheet used in a vehicle check-up, checking and finishing all the work that has already been done throughout the maintenance process.

7.1.6 Steps to develop and use a check sheet

(1) Clearly define which *events* (problem or process) are to be recorded. Add a category of 'other' to capture incidents not easily categorised into any of the specified groups.

Equipment	Worker	Monday am	Monday pm	Tuesday am	Tuesday pm	Wednesday am	Wednesday pm	Thursday am	Thursday pm	Friday am	Friday pm	Saturday am	Saturday pm
Machine 1	A	OOX•	OX	OOO	OXX	OOOXX X•	OOOOX XX	OOOOX ••	OXX	OOOO	OO	O	XX•
	B	OXX•	OOOXX O	OOOOO OXX	OOOXX	OOOOO OXX•	OOOOO OX•	OOOOO X	OOOX• •	OOXX•	OOOOO	OOX	OOO•X OX
Machine 2	C	OOX	OX	OO	•	OOOOO	OOOOO OX	OO	O•	OOΔ	OO■	ΔO	O■
	D	OOX	OX	OO	OOO•	OOO•Δ	OOOOO X	O•O	OOΔ	OOΔΔ	O••	■OOX	XXO

O Surface scratch   Δ Defective finishing   ■ Other
X Blowhole   • Improper shape

Figure 7.4.   Defective cause check sheet.

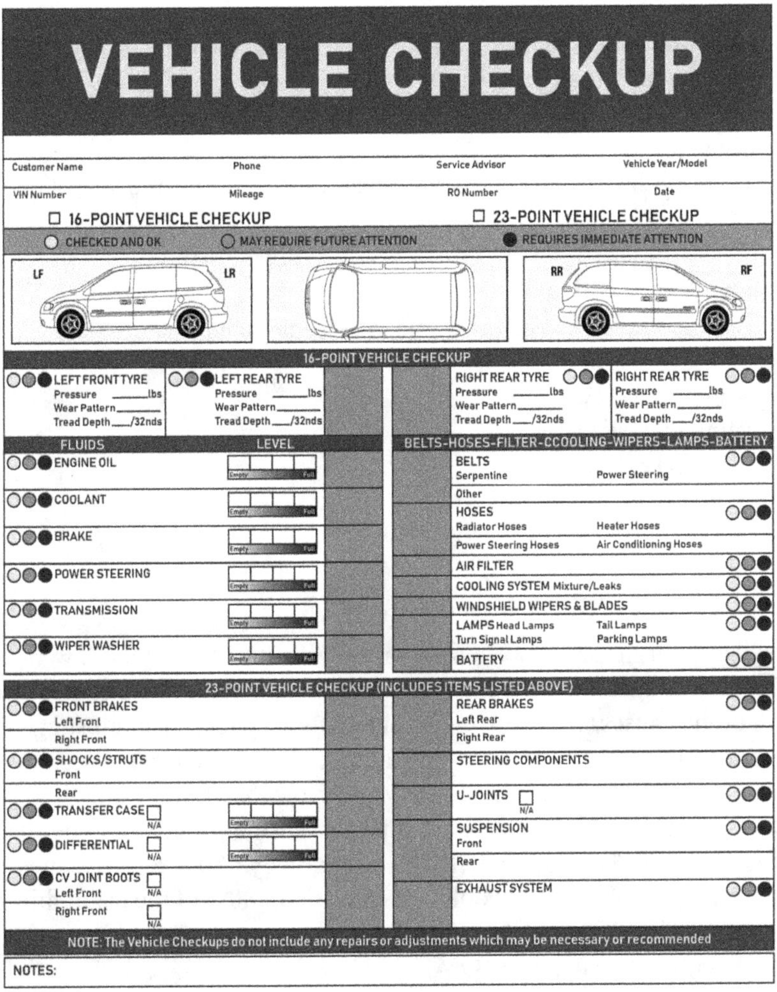

VEHICLE CHECKUP

Customer Name | Phone | Service Advisor | Vehicle Year/Model

VIN Number | Mileage | RO Number | Date

☐ 16-POINT VEHICLE CHECKUP ☐ 23-POINT VEHICLE CHECKUP

○ CHECKED AND OK ○ MAY REQUIRE FUTURE ATTENTION ● REQUIRES IMMEDIATE ATTENTION

LF LR RR RF

16-POINT VEHICLE CHECKUP

LEFT FRONT TYRE Pressure ____lbs Wear Pattern____ Tread Depth___/32nds

LEFT REAR TYRE Pressure ____lbs Wear Pattern____ Tread Depth___/32nds

RIGHT REAR TYRE Pressure ____lbs Wear Pattern____ Tread Depth___/32nds

RIGHT REAR TYRE Pressure ____lbs Wear Pattern____ Tread Depth___/32nds

FLUIDS | LEVEL

ENGINE OIL (Empty–Full)

COOLANT (Empty–Full)

BRAKE (Empty–Full)

POWER STEERING (Empty–Full)

TRANSMISSION (Empty–Full)

WIPER WASHER (Empty–Full)

BELTS-HOSES-FILTER-CCOOLING-WIPERS-LAMPS-BATTERY

BELTS Serpentine, Power Steering, Other

HOSES Radiator Hoses, Heater Hoses, Power Steering Hoses, Air Conditioning Hoses

AIR FILTER

COOLING SYSTEM Mixture/Leaks

WINDSHIELD WIPERS & BLADES

LAMPS Head Lamps, Tail Lamps, Turn Signal Lamps, Parking Lamps

BATTERY

23-POINT VEHICLE CHECKUP (INCLUDES ITEMS LISTED ABOVE)

FRONT BRAKES Left Front, Right Front

SHOCKS/STRUTS Front, Rear

TRANSFER CASE ☐ N/A (Empty–Full)

DIFFERENTIAL ☐ N/A (Empty–Full)

CV JOINT BOOTS Left Front ☐ N/A, Right Front ☐ N/A

REAR BRAKES Left Rear, Right Rear

STEERING COMPONENTS

U-JOINTS ☐ N/A

SUSPENSION Front, Rear

EXHAUST SYSTEM

NOTE: The Vehicle Checkups do not include any repairs or adjustments which may be necessary or recommended

NOTES:

Figure 7.5. Check-up confirmation check sheet (checklist).

(2) Define the *period* for data recording and suitable intervals. The time period should be representative (that is, a one-day sample on Monday may not be representative of a typical day).

(3) *Design* the check sheet to be used during data recording, allocating space for recording and for summarising within the intervals and the entire recording period.

(4) Develop a check sheet that is *easy to understand*. All columns should be clearly labelled.

(5) Perform the data collection during the agreed period, ensuring that everyone understands the tasks and the events to be recorded. Plot the information on a check sheet.

(6) *Analyse* the data to identify events with unusually high or low occurrences.

(7) Train all those involved on gathering data. A uniform data collection technique is vital.

Activity 7.1:

Which type of check sheet(s) would you recommend for:

- Testing TV sets
- Testing material hardness
- Testing a large casting
- Car MOT

7.2 Pareto Analysis

The Pareto principle (also known as the '80–20 rule', or the law of the '*vital few*') states that, for many events, 80% of the effects comes from 20% of the causes. Business management thinker Joseph Juran suggested the principle and named it after the Italian economist Vilfredo Pareto, who observed that 80% of income in Italy went to 20% of the population. It is a common rule of thumb. For example, in business, '80% of your sales comes from 20% of your clients'; in criminology, '80% of the crimes are committed by 20% of the people'; in education, '80% of the exam failures are due to 20% of the students'; etc.

In quality management (QM):

- in any population contributing to a common effect, a relatively small proportion of the contributors—the vital few—account for the bulk of the effect;
- 20% of the causes in any problem are responsible for 80% of the defects (20/80 criterion).

An important tool in Pareto analysis is the Pareto diagram (chart). A Pareto diagram is a form of bar chart where the values plotted are arranged in descending order. Figure 7.6 shows an example of Pareto diagram representing

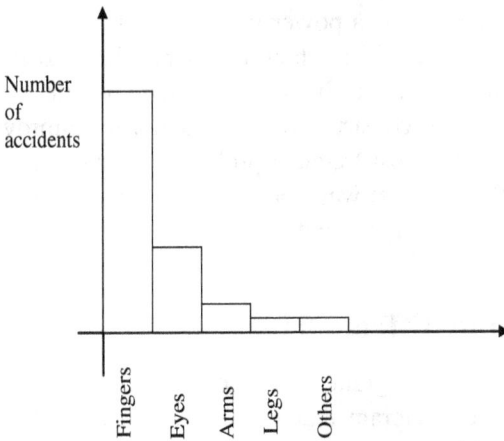

Figure 7.6. Pareto diagram for injuries.

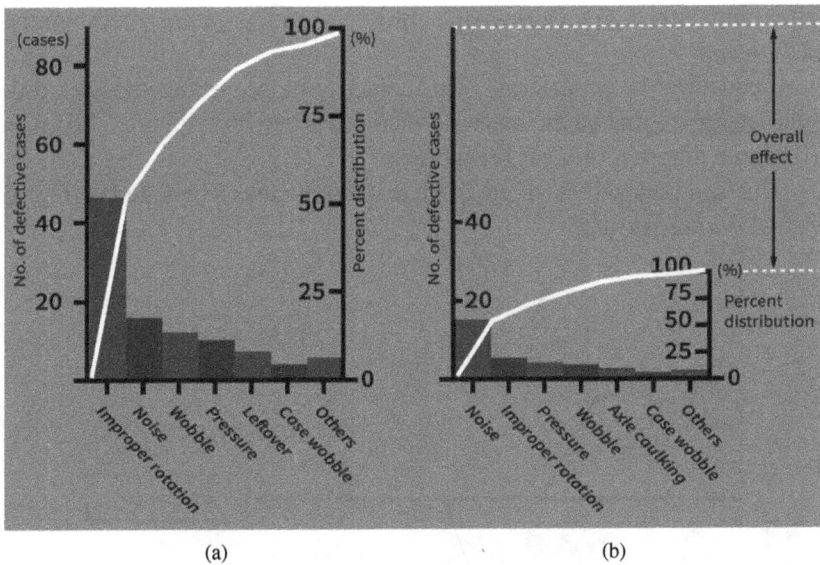

Figure 7.7. Pareto diagram (a) before improvement and (b) after improvement.

work injuries. Often, the graph is accompanied by a line graph which shows the percentage distribution of each category from left to right and the cumulative totals (see Figure 7.7(a)).

The Pareto diagram is a powerful tool for quality improvement as it identifies and helps prioritise which major problems to fight in order to get maximum improvement. Figure 7.7(b) demonstrates the effect of the use of Pareto diagram constructed before and after improvements focusing the efforts on the most frequent problem (improper rotation). We can see that while the problem was not completely eliminated, being significantly reduced yielded a massive overall effect.

7.3 Cause-and-Effect Diagram

The cause-and-effect diagram, also known as the 'fishbone diagram' is simply a structured diagram that shows the causes and sub-causes of a certain event/problem. The general structure of a cause-and-effect diagram is shown in Figure 7.8.

The cause-and-effect diagram can be used as a tool to sort out and better understand the causes of a problem and organise their mutual relationship.

Frequently, in engineering, the main causes are grouped under four main categories: machines, materials, methods and people. Figure 7.9 depicts a fishbone diagram of a car breakdown.

This method/tool is often used to record ideas generated during a *brainstorming session*.

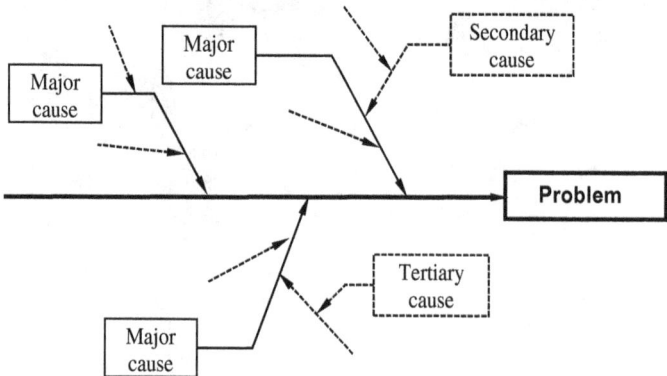

Figure 7.8.   General form of a cause-and-effect diagram.

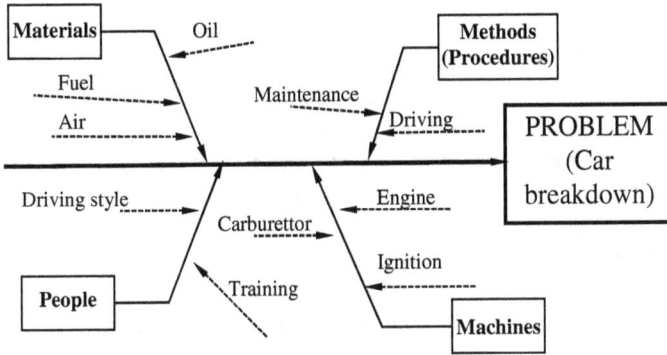

Figure 7.9. Fishbone diagram of a car breakdown.

7.3.1 *How to generate a cause-and-effect diagram?*

1. Gather a team of skilful people related to the problem for brainstorming.
2. State the problem, considering the key factors (major causes) affecting the issue.
3. Brainstorm on the question: 'What could cause this problem?'.
4. Having identified sub-causes, the question can be repeatedly asked to determine possible causes of each sub-cause. Usually, the process is stopped when the root cause of the sub-cause is found.

It should be emphasised again that the success of a brainstorming session depends on the people who are selected to be questioned in the session. If the selection of the team is not quite appropriate, the result might be a poor fishbone diagram, as shown in Figure 7.10, having only major causes and not being able to go into the true depth of the problem.

Activity 7.2:
Make a layout for a comprehensive cause-and-effect diagram on the problem, 'a student failed an exam'.

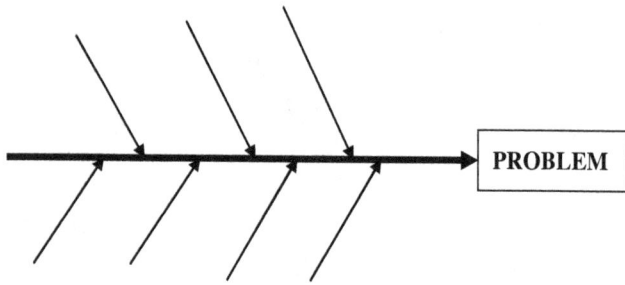

Figure 7.10. Example of a poor fishbone diagram.

7.4 Data Correlation

Many QM problems require an estimation of the relationship between two or more variables, e.g. machining time and tool wear, machine tool load and power consumption, nozzle temperature and delamination defects in 3-D printing. The correlation analysis is a powerful tool for estimating the relationship between two separate causes. A high level of correlation means that two variables are kind of dependant, meaning that if you change one, the other will change accordingly. Therefore, if you know that two parameters are somehow correlated, then you can indirectly control a parameter by controlling another.

Table 7.1 shows the results of a study on tool life Y of four identical cutting tools running at four discrete levels of feed rate X (90, 100, 105 and 110).

A data plot, known as scatter diagram (Figure 7.11), could be used for a fast visual assessment of the correlation between two variables.

Looking at the scatter diagram above, we could assume a linear relationship (linear correlation) between the variables, and with a simple graphical approach, we can try to fit a straight-line model between the points (Figure 7.12).

Using this simplified graphical approach, we can easily answer the question: if you want to keep the tool life around 40 min, what feed rate should be used?

A much more accurate and reliable estimation of the relationship between two datasets (X and Y) than with a diagram is the *correlation coefficient* r, which can be computed as a ratio between the

Table 7.1. Tool life Y vs. feed rate X.

Tool No.	Feed Rate, mm/min			
	90	**100**	**105**	**110**
	Tool Life, min			
1	41	22	21	15
2	43	35	13	11
3	35	29	18	6
4	32	18	20	10

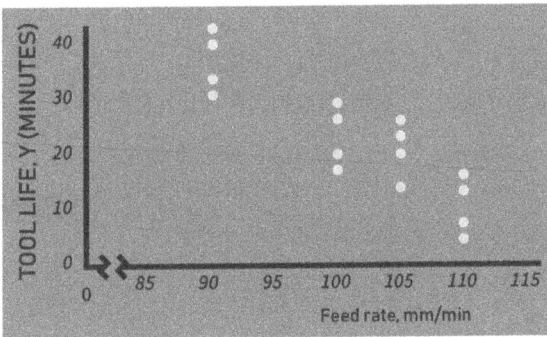

Figure 7.11. Tool life Y vs. feed rate X.

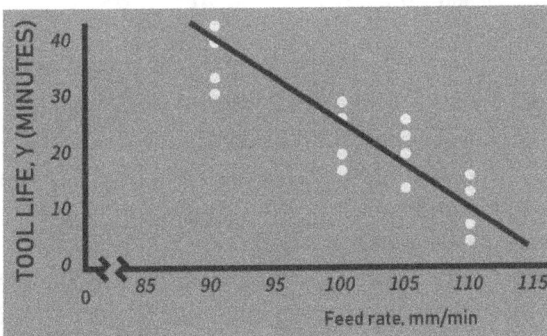

Figure 7.12. Scatter diagram with fitted straight-line model.

co-variation (cov) and the product of the standard deviations of the two variables:

$$r_{xy} = \frac{cov(X,Y)}{S_x.S_y}, \tag{7.1}$$

where

$$cov(X,Y) = \frac{1}{n-1}\sum_{i=1}^{n}(X_i - \bar{X}).(Y_i - \bar{Y}) \quad \text{and} \tag{7.2}$$

S_x and S_y are the standard deviations.

7.4.1 *Typical values of the correlation coefficient*

Figure 7.13 shows the five most typical values of data correlation and their corresponding graphical representations (scatter diagrams).

7.4.2 *Data correlation and stratification*

A proper data correlation analysis may require data stratification. Although its definition is seemingly straightforward, stratification is a term that can be used to characterise either the design of a study (e.g. stratified sampling) or alternatively, an analytic approach (stratified analysis) that can be applied to data which have previously been collected. In both cases, stratification is used because the study population consists of sub-populations or sub-domains that are of particular interest to the researcher.

Example: A manufacturing team drew a scatter diagram to test whether product purity and iron contamination were related, but the plot (Figure 7.14) did not demonstrate a relationship.

Then, a team member realised that the data came from three different reactors. The team member redrew the diagram (see Figure 7.15) using a different symbol for each reactor's data:

Now, patterns can be seen. The data from reactor 2 and reactor 3 are circled. Even without doing any calculations, it is clear that for these two reactors, purity decreases as iron increases. However, the data from reactor 1, the solid dots that are not circled, do not show this. Something is different about reactor 1.

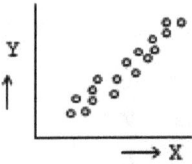

1) **Strong positive correlation ($r_{x,y} \approx 1$)**

An increase in Y depends on increases in X.

If X is controlled, Y will be naturally controlled.

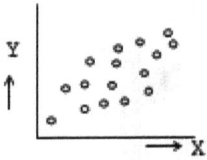

2) **Positive correlation ($0 < r_{x,y} < 1$)**

If X is increased, Y will increase somewhat, but Y

seems to have causes other than X.

3) **No correlation at all ($r_{x,y} \approx 0$)**

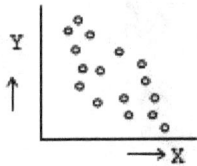

4) **Negative correlation ($-1 < r_{x,y} < 0$)**

An increase in X will cause tendency for decrease in Y.

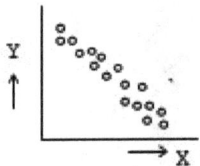

5) **Strong negative correlation ($r_{x,y} \approx -1$)**

An increase in X will cause a decrease of Y.

X may be controlled instead of Y.

Figure 7.13.   Various types of data correlations.

Here are examples of different sources that might require data to be stratified:

- Equipment
- Shifts

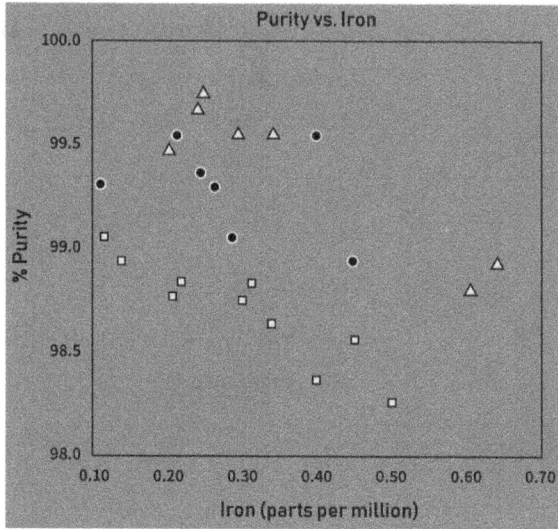

Figure 7.14.   Purity vs. Iron, no correlation.

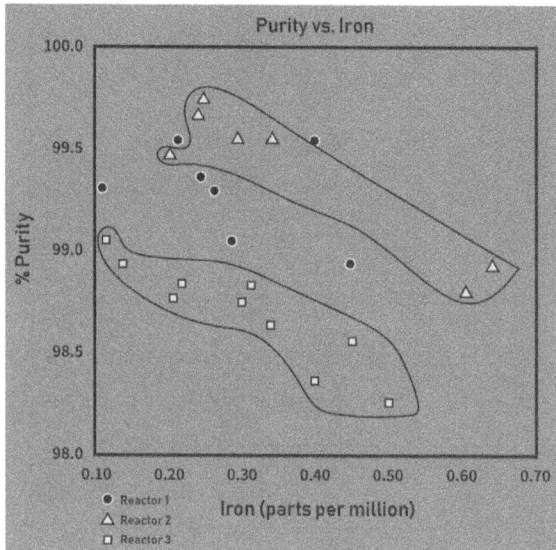

Figure 7.15.   Purity vs. Iron, after data stratification, showing strong negative correlation.

- Departments
- Materials
- Suppliers
- Day of the week
- Time of day
- Products

Survey data usually benefit from stratification. Some important data stratification considerations are as follows:

- Before collecting data, consider whether stratification might be needed during analysis. Plan to collect stratification information. After the data are collected, it might be too late.
- On your graph or chart, include a legend that identifies the marks or colours used.

7.5 Regression Analysis

Regression analysis is a statistical technique used for optimisation and prediction in QM. Regression analysis uses the so-called 'black box' approach. In science, computing and engineering, a black box is a device, system or object, which can be viewed only in terms of its inputs and outputs, without any knowledge of its internal workings or mechanism.

The inputs are made by a set of independent variables $X_1 \ldots X_n$ affecting one way or another the output represented by a dependent variable, Y (Figure 7.16).

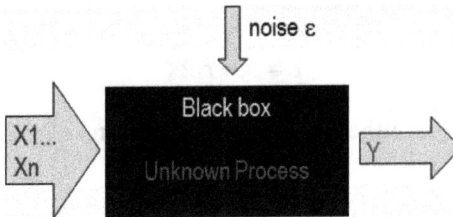

Figure 7.16. 'Black box' approach in regression analysis.

Usually, the independent variables can be controlled or monitored. Any variables that cannot be controlled will be regarded as 'noise', and they would affect the dependable variable Y in an unpredictable way.

Regression analysis estimates the parameters of an equation relating independent variables (regressors) to a dependent variable (regressand). The resulting equation is called the regression equation (model). Usually, the regression models propose that the regressand Y is a function of the regressors X_i and unknown coefficients β_i:

$$Y = f(X_i, \beta_i) + \varepsilon,$$

where ε is the experimental error.

The objective is to find estimates (b_0, b_1 ... b_k) of the unknown parameters (β_0, β_1 ... β_k) and thus obtain a prediction equation:

$$\hat{Y} = f(X_i, b_i),$$

where $\hat{Y}$ is the prediction value of Y (regressand) for the given values of regressors $X_1, X_2,..., X_k$.

7.5.1 *Regression models*

In order to describe the nature of the process adequately, the analysis may require a different model/equation. Complicated models can lead to very time-consuming and complex calculations. For this reason, the most popular models are linear in the coefficients. Starting with the simplest, they are:

- simple linear regression model (single regressor):

$$\hat{Y} = b_0 + b_1 X, \tag{7.3}$$

where b_0 and b_1 are the unknown intercept and slope, respectively, of the regression line;

- linear (more than one regressor):

$$\hat{Y} = b_0 + b_1 X_1 + ... + b_k X_k; \tag{7.4}$$

- quadratic (only in the independent variables), but still linear in coefficients, including mixed members, e.g. X_1X_2:

$$\hat{Y} = b_0 + b_1 X_1 + b_2 X_2 + b_3 X_1^2 + b_4 X_2^2 + b_5 X_1 X_2; \quad (7.5)$$

- nonlinear (e.g. cubic, exponential) in coefficient models are usually converted into linear in coefficients.

Sometimes, one of the regressors can be a nonlinear function of another regressor of the data. However, the model remains linear as long as it is linear in the coefficient β.

In the example given in Figure 7.12, the line was fitted graphically which can be quite subjective. 'Least-squares' method provides an analytical tool for finding estimates of these parameters from a set of n observations (Y_1, X_1), ..., (Y_n, X_n). The estimates are called 'least-squares estimates' because they minimise the sum of the squared deviations between the observed and predicted values of the regressand, which in a liner model, takes the following form:

$$\sum_{m=1}^{n} \left(Y_m - \overline{Y_m}\right)^2 = \sum_{m=1}^{n} \left(Y_m - b_0 - b_1 X_m\right)^2. \quad (7.6)$$

The least-squares estimates for the parameters of the simple linear model are

$$b_1 = \frac{\sum \left(X_m - \overline{X}\right)\left(Y_m - \overline{Y}\right)}{\sum \left(X_m - \overline{X}\right)^2}, \quad (7.7)$$

$$b_0 = \overline{Y} - b_1 \overline{X}. \quad (7.8)$$

So, assuming a simple linear model ($Y = b_0 + b_1 X$) and using the above formulae and data from Table 7.1, we can calculate the equation parameters. The calculations can be carried out using a table or a spreadsheet, as shown in Table 7.2, where b_1 is equal to the sum of the values in column 6 divided by the sum of the values in column 4:

$$b_1 = -1191.25/875 = -1.36.$$

Quality Management Essentials

Table 7.2. Manual calculation of the regression coefficients.

1	2	3	4	5	6	7
					(Xi-AverX)	
X	Y	Xi-AverX	(Xi-AverX)^2	Yi-AverY	(Yi-AverY)	(Yi-AverY)^2
90	43	−11.25	126.56	19.94	−224.30	397.50
90	41	−11.25	126.56	17.94	−201.80	321.75
90	35	−11.25	126.56	11.94	−134.30	142.50
90	32	−11.25	126.56	8.94	−100.55	79.88
100	35	−1.25	1.56	11.94	−14.92	142.50
100	29	−1.25	1.56	5.94	−7.42	35.25
100	22	−1.25	1.56	−1.06	1.33	1.13
100	18	−1.25	1.56	−5.06	6.33	25.63
105	21	3.75	14.06	−2.06	−7.73	4.25
105	20	3.75	14.06	−3.06	−11.48	9.38
105	18	3.75	14.06	−5.06	−18.98	25.63
105	13	3.75	14.06	−10.06	−37.73	101.25
110	15	8.75	76.56	−8.06	−70.55	65.00
110	11	8.75	76.56	−12.06	−105.55	145.50
110	10	8.75	76.56	−13.06	−114.30	170.63
110	6	8.75	76.56	−17.06	−149.30	291.13
Aver X	**Aver Y**	**Sum =**	**875.00**		**−1191.25**	**1958.94**
101.25	**23.06**					

Then, b_0 is calculated using the averages of X and Y:

$b_0 = \text{Aver}Y - b_1 \times \text{Aver}X = 23.06 - (-1.36) \times 101.25 = 160.91$.

So finally, the model equation will take the form

$$Y = 160.91 - 1.36X.$$

Using the above equation, we can calculate and predict the value of the tool life Y for any depth of cut X (within the range of the test results) without need of any extra tests. This equation is particularly useful if we wish a tool life to achieve a certain amount of time to guarantee that the tool will last long enough to complete the cut without breaking down and damaging the workpiece, which might be very essential.

7.5.2 *Assessment of the regression models*

Choosing a suitable model is not always an easy job and, in some cases, may lead to an equation which does not appropriately fit the process. The adequacy of the model can be assessed by 'analysis of the residuals', i.e. the difference (lack of fit) between the observed and estimated values:

$$e_m = Y_m - \hat{Y}_m. \qquad (7.9)$$

A measure of how well the regression equation fits the data is the so-called *coefficient of determination* r^2. In fact, r^2 is the square of the *correlation coefficient*, r, and it represents the 'goodness of fit', ranging from zero to one.

For example, the following figure shows the relationship between cutting feed X and surface finish Y. A linear regression model has been suggested and a straight line has been fitted using 'least squares' (see Figure 7.17).

Studying and plotting the residuals for all observations, we may see one of the following pictures:

- Random pattern (Figure 7.18(a)): The coefficient of determination r^2 is close to one, which means that the regression model is adequate and could be used for prediction and optimisation.

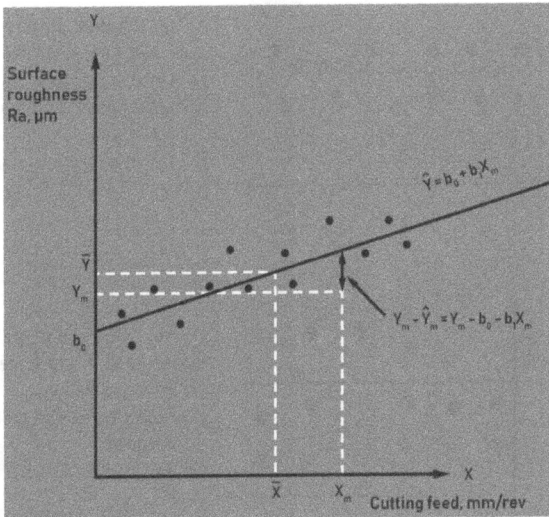

Figure 7.17.   Model line fitted using 'least squares' method.

- Non-random pattern (Figure 7.18(b)): If the coefficient of determination r^2 is close to zero, the regression equation is not helpful in predicting a y value as there is a lack of correlation between the estimated and the actual values.

An r^2 of 0.10 means that only 10% of the variance in Y is predictable from X, an r^2 of 0.20 means that 20% is predictable and so on.

If a regression model is found to be inadequate, an alternative model should be tested.

A regression model having low coefficients of determination could be improved by adding an extra member to the equation.

7.5.3 *Multiple regression*

We can analyse how a single dependent variable is affected by the values of one or more independent variables, e.g. how an athlete's performance is affected by factors such as age, height and muscle mass. We can apportion shares in the performance measure to each of these three factors, based on a set of performance data, and then use the results to predict the performance of a new, untested athlete. Unfortunately, a higher number of

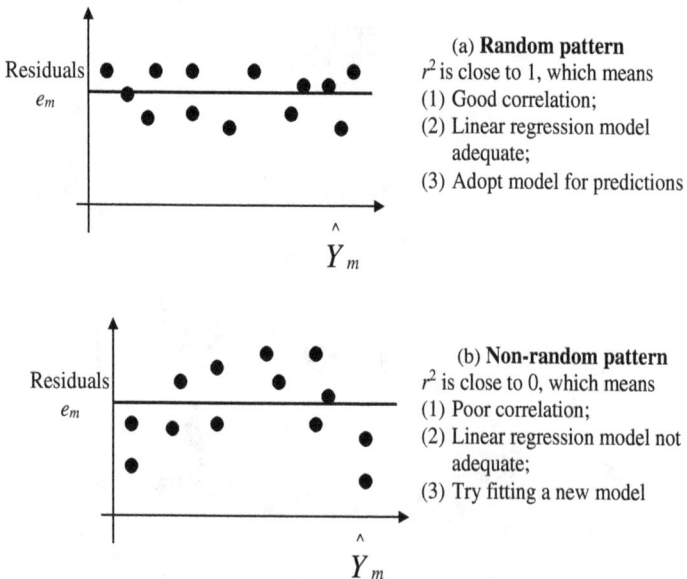

(a) **Random pattern**
r^2 is close to 1, which means
(1) Good correlation;
(2) Linear regression model adequate;
(3) Adopt model for predictions

(b) **Non-random pattern**
r^2 is close to 0, which means
(1) Poor correlation;
(2) Linear regression model not adequate;
(3) Try fitting a new model

Figure 7.18. Random pattern (a) and non-random pattern (b).

independent variables would dramatically increase the complexity of the calculations, with a larger number of coefficients/parameters to be worked out and more tests needing to be done.

$$\hat{Y} = b_0 + b_1 X_1 + b_2 X_2 + b_3 X_1^2 + b_4 X_2^2 + b_5 X_1 X_2, \qquad (7.10)$$

where the product $X_1 X_2$ is a combination of two factors (e.g. height and muscle mass).

It is important to note that the number of observations **n** must be greater than the number of members/parameters of the model $b_0, b_1 \ldots b_m$.

In addition, when selecting the set of independent variables, we must make sure that they are not correlated. For example, if we study the fuel consumption of a vehicle, we should not include load and the overall weight of the vehicle as they are correlated.

Warning: With the calculation of predictions based on a regression model, do not use values for the independent variable that are outside the range of values used to create the equation. Often, such extrapolation can produce unreliable estimates.

Regression analysis example:
Problem statement:

Last year, five randomly selected students took a math's test before they began their course in engineering. The engineering department has three questions:

1. Can you find a model/equation to predict course performance based on math's test scores?
2. Estimate how good this model is for predictions.
3. If a student made an 80 on the math's test, what course grade would we expect him/her to achieve?

The department provided data organised in Table 7.3, where X_i is the score on maths test in % and Y_i is the course grade in %.

Answer:
Step 1. The regression model is a linear equation of the form $\hat{y} = b_0 + b_1 X$. To conduct a regression analysis, we need to solve for b_0 and b_1. Computations are shown in Table 7.4.

Table 7.3. Performance data set.

Students	X_i	Y_i
1	95	85
2	85	95
3	80	70
4	70	65
5	60	70

Table 7.4. Computation table.

Student No.	X_i	Y_i	$(X_i - \bar{X})$	$(Y_i - \bar{Y})$	$(X_i - \bar{X})^2$	$(Y_i - \bar{Y})^2$	$(X_i - \bar{X})*(Y_i - \bar{Y})$
1	95	85	17	8	289	64	136
2	85	95	7	18	49	324	126
3	80	70	2	−7	4	49	−14
4	70	65	−8	−12	64	144	96
5	60	70	−18	−7	324	49	126
Sum	390	385					
Average	78	77					

So, the parameters b_0 and b_1 can be calculated as follows:

$$b_1 = \frac{\sum (X_m - \bar{X})(Y_m - \bar{Y})}{\sum (X_m - \bar{X})^2} = 470/730 = 0.644,$$

$$b_0 = \bar{Y} - b_1 \bar{X} = 77 - (0.644) \times (78) = 26.768.$$

Thus, the model equation will take the form

$$\hat{y} = 26.768 + 0.644x.$$

Step 2. To find the correlation coefficient r_{xy}, the formula (7.1) is combined with (7.2) to take the following form:

$$r_{xy} = \frac{\sum_i (X_i - \bar{X})(Y_i - \bar{Y})}{\sqrt{\sum_i (X_i - \bar{X})^2} \sqrt{\sum_i (Y_i - \bar{Y})^2}}.$$

(7.11)

In this case, we can calculate r_{xy} using the results from Table 7.4 as follows:

$$r_{xy} = \frac{470}{\sqrt{730}\sqrt{630}} = 0.693.$$

The coefficient of determination would be

$$r^2 = 0.693^2 = 0.480.$$

An r^2 of 0.48 means that 48% of the variance in Y is predictable from X. The remaining 52% would depend on different factors.

Step 3. Using the model, we can calculate the predicted grade:

$$\hat{y} = 26.768 + 0.644x$$

$$= 26.768 + 0.644 \times 80 = 26.768 + 51.52 = 78.288.$$

The graphical representation would be as follows (Figure 7.19):

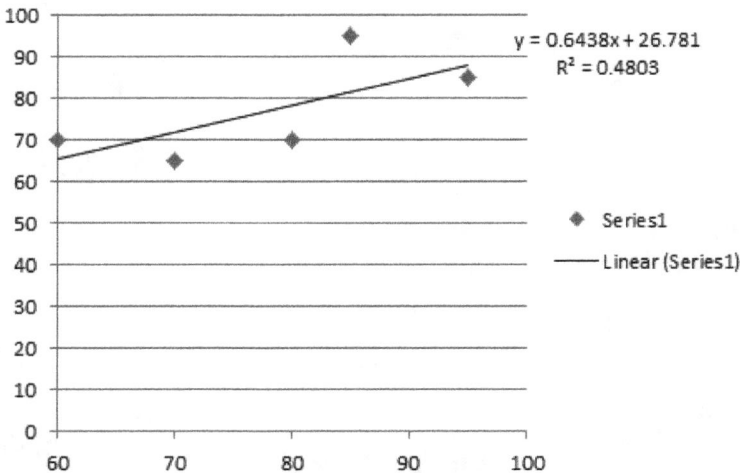

Figure 7.19.   Scatter diagram with superimposed linear model.

Activity 7.3:
Study the relationship between speed (V, miles/hour) and fuel cost (Y, pence/litre). The data presented in Table 7.5 have been obtained by experiments using three test cars (Car 1, Car 2 and Car 3) each driven at three different speeds (30, 50 and 70 miles/hour, respectively).

(1) Plot a scatter diagram. Visually and analytically study the correlation between the two parameters.
(2) Suggest an appropriate regression model for the data and work out the equation parameters.
(3) Calculate the fuel cost at 40 miles/hour. Give a graphical representation.

Table 7.5.   Speed vs. cost data.

Speed	Car 1	Car 2	Car 3
30	10	12	11
50	15	16	15
70	25	26	24

Chapter 8

Six-Sigma Quality

8.1 Is the So-called 'Satisfactory' Quality Really Good Enough?

Generally, if a parameter's distribution is normal with process capability index $C_p = T/6S = 1$, i.e. the process variation V, based on $\pm 3\sigma$ is equal to the design tolerance T, and if the process is very well centred ($\mu \equiv T_m$), only 0.27% of the output would be expected to fall outside the specified design tolerance T, see Figure 8.1.

However, consider what this level of quality really means, i.e. what if 0.27% is outside the design tolerance range? For example:

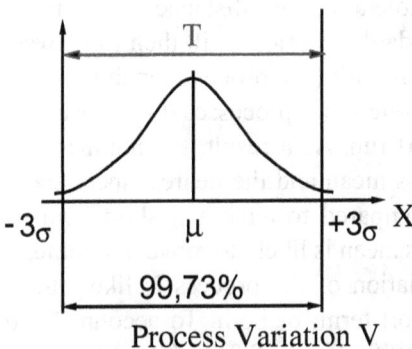

Figure 8.1. $C_p = 1$.

- at least 20,000 wrong drug prescriptions each year;
- more than 15,000 babies accidentally dropped each year by nurses and obstetricians;
- no electricity, water or heat for about hours each year;
- 500 incorrect surgical operations each week;
- 2,000 lost pieces of mail each hour.

Remember, this is the ideal case when the process is dead well centred! Besides, experience has shown that in the long term, processes do not usually perform as well as they do in the short run, so over time it may only get worse. Obviously, in many cases, this is NOT quite an acceptable quality.

8.2 Origin and Meaning of the Term 'Six-Sigma Process'

Often, the standard deviation (SD, σ, sigma) is used as a measure of variation of a population. The term 'Six-Sigma process/quality' comes from the notion that if there are six standard deviations between the mean of a process and the nearest specification limit, there will be practically no items that fail to meet the specifications. This is based on the calculation method employed in a process capability study, often used by quality professionals. The term 'Six-Sigma' has its roots in this tool. In a capability study, the number of standard deviations between the process mean and the nearest specification limit is given in 'sigma' units. As the process standard deviation goes up or the mean of the process moves away from the centre of the tolerance, the distance in sigma numbers goes down because fewer standard deviations will then fit between the mean and the nearest specification limit (see process capability index C_{pk}). Experience shows, that in the long term, processes do not usually perform as well as they do in the short run. As a result, the number of sigmas that will fit between the process mean and the nearest specification limit is likely to drop over time, compared to an initial short-term study. This is either because the process mean is likely to move over time, or because the long-term standard deviation of the process is likely to be greater than that observed in the short term, or both. To account for this, an empirically based ±1.5 sigma shift (Figure 8.2) is introduced into the calculation of the quality. In other words, a process that fits six sigmas between the

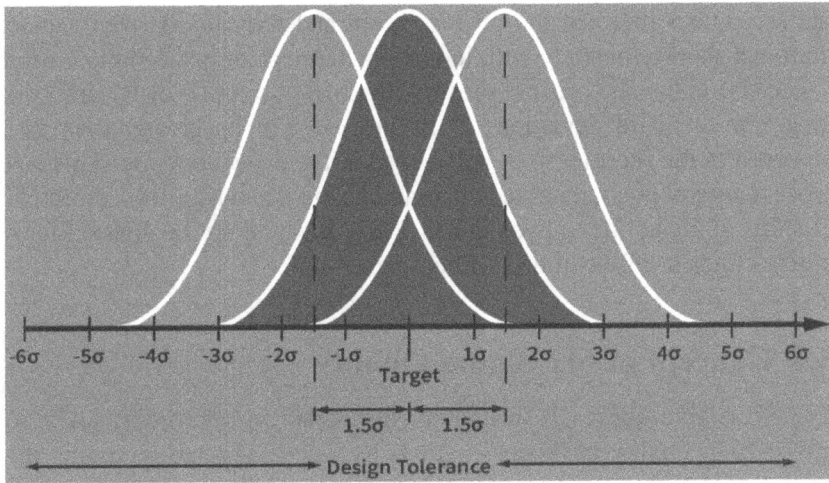

Figure 8.2. Six-Sigma concept.

process mean and the nearest specification limit in a short-term study will only fit 4.5 sigmas in the long term.

Hence, the widely accepted definition of a Six-Sigma process (quality) is one that produces only 3.4 defective parts per million opportunities (DPMO). This is based on the fact that a process which is normally distributed will have 3.4 parts per million beyond a point that is 4.5 standard deviations above or below the mean (one-sided capability study). So, the 3.4 DPMO of a 'Six-Sigma' process in fact corresponds to 4.5 sigmas, namely six sigmas minus the 1.5 sigmas shift introduced to account for long-term variation/shift. This is designed to prevent overestimation of real-life process capability. Theoretically, if the process is held exactly on the target (the shaded distribution in Figure 8.2), only 2.0 defects per billion would be expected.

8.3 Why Allow a Process Shift of 1.5 Standard Deviations?

As mentioned earlier, processes do not usually perform as well in the long term as they do in the short term as they tend to have increased variation or drift away from the target or both. According to this idea, a process that fits six sigmas between the process mean and the nearest specification limit in a short-term study will, in the long term, fit only 4.5 sigmas. This is because either the process mean will move over time, or because the long-term standard deviation of the process will be greater than that observed in the short term, or both. In addition, many common statistical process control plans are based on a sample size that only allows detection of shifts of about one to two sigmas. Thus, it would not be unusual for a process to drift off this much without being noticed.

8.4 Concept of Six-Sigma Quality

The Six-Sigma approach (originally developed by Motorola) seeks to improve the quality of process outputs by identifying and removing the causes of defects (errors) and minimising variability in manufacturing and business processes. The term 'Six-Sigma' comes from statistics and is used in statistical quality control, which evaluates process capability. Originally, it referred to the ability of manufacturing processes to produce a very high proportion of output within specification. Processes that operate with 'Six-Sigma quality' over the short term are assumed to produce long-term defect levels below 3.4 DPMO. The term 'Six-Sigma' refers to the ability of highly capable processes to produce output within specification.

8.5 Achieving Six-Sigma Quality

Six-Sigma utilises many established quality-management tools that are also used outside Six-Sigma, such as analysis of variance, regression analysis, correlation, scatter diagram, chi-squared test, cause-and-effects diagram and control charts.

In a simple statistical sense, achieving Six-Sigma requires satisfying two conditions:

(1) Shrinking the process variation to half of the design tolerance (Figure 8.3), i.e. $C_p = 2.0$ or more is required, meaning that we introduce a working tolerance of half the design tolerance.

Since, $C_p = \dfrac{T}{6\sigma} = 2$,

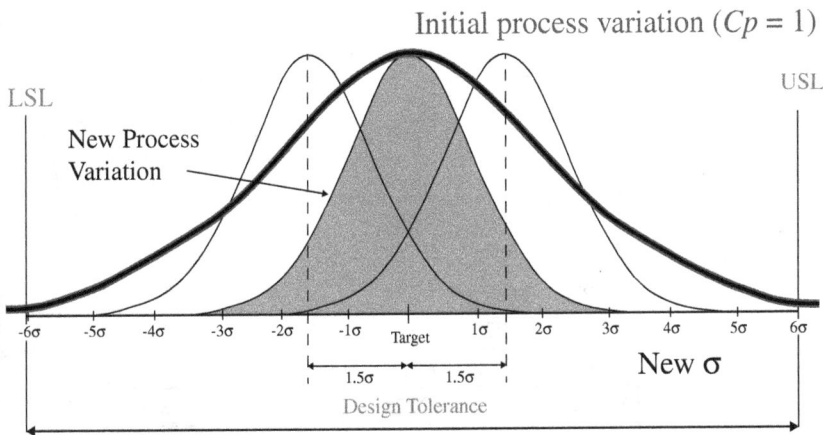

Figure 8.3.   Initial process variation shrunk to half the size.

we can calculate the required SD as $\sigma = \dfrac{T}{12}$;

(2) Allowing the mean to shift no more than $\pm 1.5\sigma$ from the target (usually the middle of the tolerance T_m).

8.5.1 *Do we always have to require minimum $\sigma = 1/12$ of tolerance T?*

A quality level of 3.4 defects per million can be achieved in several ways, for instance, five sigmas with 0.5 sigma off-centring allowed.

Table 8.1 shows the trade-offs between the allowable off-centring and quality level in terms number of sigmas.

So, in other words, if the mean's off-centring (drift) can be maintained in a tighter tolerance than 1.5 sigmas around the target (tolerance middle), then the allowed process variation could be a bit wider (i.e. a process with bigger sigma or less than six sigmas to the nearest design limit) with the process still achieving Six-Sigma quality level.

8.6 Some Criticism of Six-Sigma Quality

Is it always economically efficient to implement Six-Sigma quality? In some cases, controlling the process to stay centred on the target is less

Table 8.1. Defects per million (DPM).

Off-Centering (Sigma)	Sigma Quality Level						
	3	3.5	4	4.5	5	5.5	6
0	2700	465	63	7	0.6	0.04	0.002
0.25	3557	665	99	12	1.1	0.08	0.005
0.5	6442	1382	236	32	3.4	0.29	0.02
0.75	12313	2990	578	88	11	1.0	0.1
1	22782	6213	1350	233	32	3.4	0.3
1.25	40070	12225	2980	577	88	11	1.0
1.5	66811	22750	6210	1350	233	32	3.4
1.75	105651	40059	12224	2980	577	88	11
2	158656	66807	22750	6210	1350	233	32

expensive than reducing the process variability. Table 8.1 can help us assess the trade-offs. While 3.4 defects per million might work well for certain products/processes, it may not be ideal for others. A pacemaker may need higher standards, whereas a direct mail advertising campaign may need lower ones. The basis and justification for choosing six as the number of standard deviations is not clearly explained. Motorola, for example, has achieved Six-Sigma capabilities in some processes but only Four- or Five-Sigma levels in most others.

8.7 Meanings of Six-Sigma Quality

Nowadays, there are many books on Six-Sigma quality.

It uses a set of quality-management methods, mainly empirical and statistical methods, to create a special infrastructure of people within the organisation who are experts in these methods. Each Six-Sigma project carried out within an organisation follows a defined sequence of steps and has specific value targets, e.g. reduce process cycle time, reduce pollution, reduce costs, increase customer satisfaction and increase profits.

In a simple statistical sense, the meaning is to have 12 sigmas inside specification (tolerance), i.e. to have no more than 3.4 defects per million.

In general terms, Six-Sigma means:

- continuous efforts to reduce variation in process outputs and improvement towards 'zero defects';
- manufacturing and business processes can be measured, analysed, improved and controlled;
- succeeding at achieving sustained quality improvement requires commitment from the entire organisation, particularly from top-level management.

Activity 8.1:
If a parameter having a specified tolerance 300 ± 6 is normally distributed with SD = 1.00 and mean = 298.0, check if Six-Sigma quality has been statistically achieved or otherwise.

Chapter 9

Acceptance Sampling

Consider the following typical scenario in the industry:

You have just received a shipment of 50,000 parts from a new supplier. You are in charge of deciding whether the shipment is good enough to put into your inventory. What would you do?

Your options are as follows:

- Look at all 50,000 parts (100% inspection) — it would take a very long time.
- Don't look at any and put the whole shipment into stock (0% inspection) — too risky given this is a new supplier.
- Look at some parts (a sample), and if enough of these are good, accept the whole lot (acceptance sampling).

9.1 What is Acceptance Sampling?

Acceptance sampling (AS) is an important aspect of statistical quality control. It originated back during World War II when the military had to very quickly determine which batches of ammunition to accept and which ones to reject. They knew that they could not test every bullet to find out if it will do its job in the field. On the other hand, they had to be confident that the bullets they were getting would not fail when their lives were already on the line. AS was the answer — testing a few representative bullets from the lot, so they will know/estimate how the rest of the bullets would perform. AS can be regarded as a good compromise between not doing any inspection at all and running an 100% inspection.

9.2 AS Procedure

(1) A random sample should be picked from the batch/lot (box) (see Figure 9.1) of finished products and each item from this sample is then measured/examined/tested.
(2) A decision to accept or reject this lot is based on a comparison between the number of defectives found in the sample and the number of defectives that we still consider acceptable for that sample size.

An important point to remember is that *the main purpose of AS is to decide whether or not a lot is likely to be acceptable, not to estimate the quality of the lot.*

9.3 Some Terminology of Acceptance Sampling

- *Defect*: The non-fulfilment of an intended usage requirement. Example: a crack in a glass bottle.
- *Defective (non-conforming) item*: A unit of production, which has at least a single essential defect.
  Example: a glass bottle having a crack is deemed to be defective.

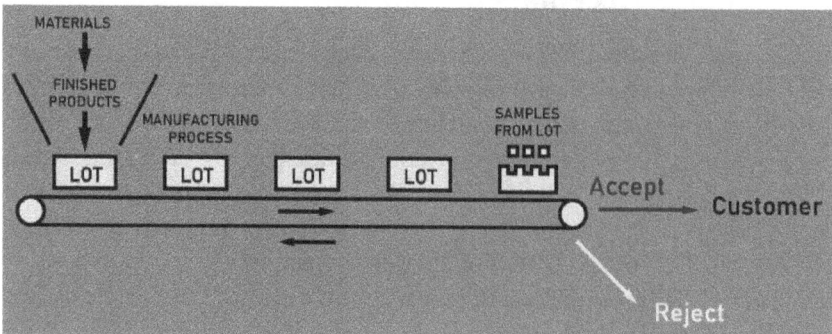

Figure 9.1.   (Lot) acceptance sampling — the core of sampling inspection.

- *Non-conformity*: The non-fulfilment of a specified requirement. Example: a 1-litre glass bottle having an actual volume of only 0.9 litre.

9.4 What Does the Measurement of a Sample Tell Us?

- Whether the pieces in the sample are good or bad (based on facts/results of measurements);
- Whether the process, at the time the sample was made, was doing good work or bad (assumption);
- Whether the uninspected pieces in the lot are good or bad (principle of sample inspection);
- Whether pieces currently unmanufactured are going to be good or bad (extrapolated assumption).

9.5 Problems of Total Inspection (100% Inspection)

9.5.1 *Cases where total inspection is not practical*

- It is costly.
- It may take too long.
- It is not always necessary (e.g. a bag of cheap nails).
- Inspection may cause damage or even complete destruction of the items (handling delicate products, such as chocolate sweets, cherries and berries).
- Inspection may be a hazardous and even a dangerous procedure (molten metal, harmful gas, etc.).

9.5.2 *Situations where inspection by sampling is indispensable*

- Destructive testing (ammunition such as bullets and shells, tasting food, drinks, etc.).
- Inspecting long lengths/large volumes of goods (copper coils, photographic films, paper, textile, thread, cement, sand, etc.).
- Inspecting very large amounts (nuts, bolts, seeds, grains, etc.).

9.6 Cases Where Inspection by AS is Not Applicable

Basically, there are situations where any risk in production is not acceptable. Usually, these are products or services related to health and safety (e.g. checking parachutes, car brakes, safety belts), products to be sent to space, etc.

9.6.1 *Type of risks in AS*

With AS (unlike 100% inspection), the actual number of defective items/ defects in the lot is unknown. It is predicted on the basis of the sample taken. Therefore, the following risks apply:

- *Producer's risk*: to reject a 'good' lot (error type I) because of an inaccurate prediction. A 'good' lot is a lot in which the actual number of defective items/defects does not exceed the number considered to be acceptable.
- *Consumer's risk*: to accept a 'bad' lot (error type II). A 'bad' lot is a lot in which the actual number of defective items/defects is greater than the acceptable value.

9.7 Advantages and Disadvantages of AS

9.7.1 *Advantages of AS*

- Economic — it saves time and money.
- Less handling damages.
- Applicability to destructive testing.

9.7.2 *Disadvantages of AS*

- A sample provides less information than 100% inspection, hence there are risks of accepting 'bad' lots and rejecting 'good' lots.
- There is additional planning and documentation.

9.8 Sampling Plans

AS has to be well planned and organised. An acceptance or *sampling plan* is a process in which the representative samples will be selected from a population. Then, the sample will be tested and judged against predefined criteria, which allows the determination of whether the lot is acceptable or not.

In order to design a sampling plan, we need to know:

- How many items to inspect or test?
- How many 'good' ones are enough to accept?
- How many 'bad' ones are enough to reject?

In production, AS is usually organised on the basis of the so-called *standard sampling plans*. Each plan consists of a set of rules and numbers used for decision making. These plans have been initially standardised by American military standards (e.g. MIL STD 105). Nowadays, they are described by international standards [7,8].

There are two major classifications of acceptance plans: by attributes ('GO, NOT-GO', good or bad) and by variables. The attribute case is the most common for AS and will be assumed for the rest of this chapter.

According to the number of samples taken, the sample plans fall into the following categories.

9.8.1 *Single sampling plan*

One random sample of n items is selected from a lot (batch) of N (e.g. 10,000) items, and the disposition of the entire lot is determined from the resulting information. These plans are usually denoted as (n, c) plans for a sample size n (e.g. 10), where the lot is:

- rejected if the number of defectives z found in the sample is more than the acceptable number of defectives c (e.g. 2) or
- accepted if it (z) is less than or equal to c.

The general scheme (layout) of a single sampling plan is shown in Figure 9.2.

In the figure, Ac is the acceptance number, $Ac = c$ and Re is the rejection number, $Re = c + 1$.

Note that in normal inspection, according to this plan, Ac and Re are two consecutive numbers, so the only possible outcome is a decision to accept or reject the lot. These are the most common plans as they are the easiest to understand and run. However, they are not really the most efficient in terms of the average number of samples needed.

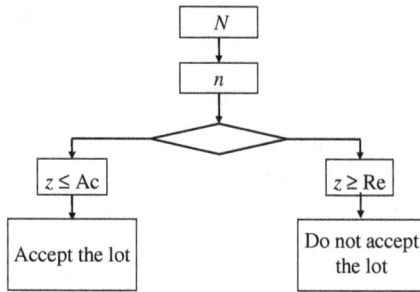

Figure 9.2.   Single sample plan scheme.

9.8.2 *Double sampling plan*

As in the first stage of this plan, Ac and Re are not consecutive numbers, once the first sample size n_1 (usually smaller than that in single sampling plan) is tested and the number of defectives z_1 is found, there are three possibilities:

(1) accept the lot if $z_1 <=$ Ac$_1$;
(2) reject the lot if $z_1 >=$ Re$_1$;
(3) no decision is possible at this stage if Ac$_1 < z_1 <$ Re$_1$.

If the outcome is the least likely possibility (3), only then at this second stage a second sample n_2 is taken, in which the number of defectives found in it is z_2. Then, the procedure (see Figure 9.3) is to combine the results $(z_1 + z_2)$ of both the samples and to make a final decision based on that information against a new set of (this time consecutive) acceptance/rejection numbers (Ac$_2$, Re$_2$).

Note that in most cases using a double sampling plan, the decision to accept or reject a lot is taken at the first stage and there is no need for a second sample.

9.8.3 *Multiple sampling plan*

The multiple sampling plan can be considered as an extension of the double sampling plan, in which more than two samples are needed to reach a conclusion (see Figure 9.4). The advantage of multiple sampling is an even smaller first sample size.

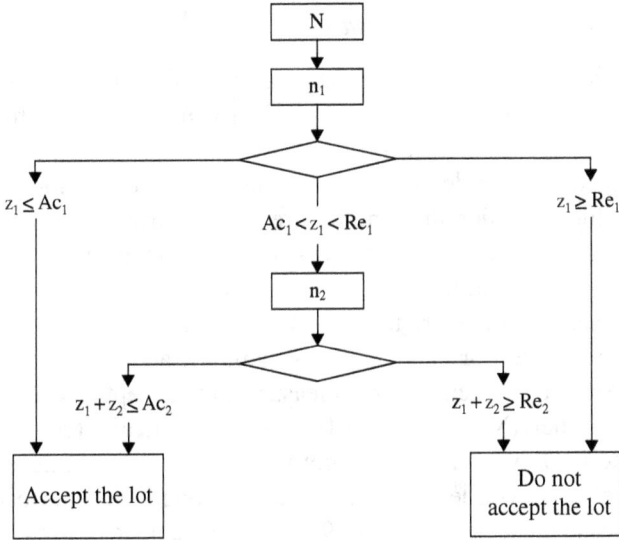

Figure 9.3. Double sampling plan.

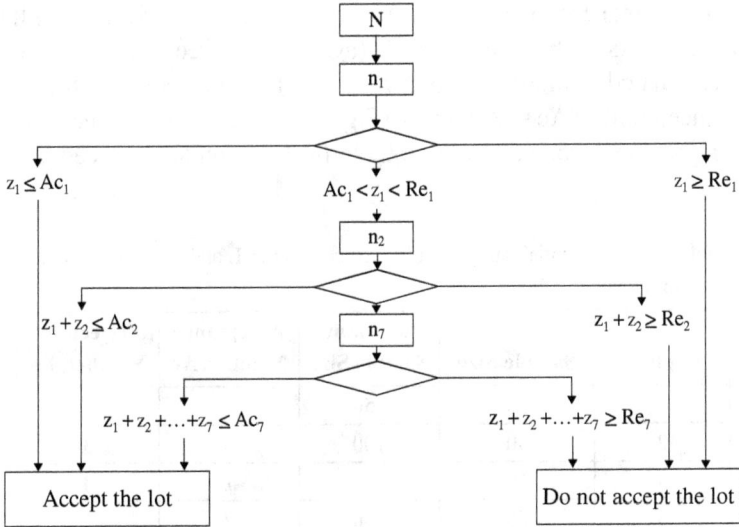

Figure 9.4. Multiple sampling plan.

9.8.4 *Sequential sampling plan*

Sequential acceptance sampling plans are the most statistically efficient type of sampling plan. They are used when you do not want to take any more sample items than necessary, but you want to be confident of having enough data to make a decision. This is the ultimate extension of the multiple sampling, in which the samples are selected from a lot one at a time and after inspection of each item, a decision is made to accept or reject the lot or select another unit.

Procedure: Samples are taken in sequence until it is possible to sentence the lot as good or bad. As you accumulate and increase the sample size with each subsequent sample measured, the acceptance and rejection numbers will increase as well, see Table 9.1. Graphically (see Figure 9.5), they tend to form two parallel straight lines, i.e. a zone. Inside is the indecision zone. Once you leave it, the lot can be judged good or bad.

In the case presented in Figure 9.5, only one defective item is found in the first sample of 50, so a decision cannot be taken. In the second sample of 50, another defective item is found, so with the accumulated number of defectives $z = 2$, again, a decision cannot be taken as $0 = Ac < z < Re = 3$.

In the third sample of 50 (150 all together), no defective items have been found (total $z = 2$), but as $1 = Ac < z = 2 < Re = 4$, a decision still cannot be taken. As in the fourth sample (again found free of defective items) the accumulated number of defectives $z = 2$ becomes equal to the current acceptance number $Ac = 2$, point 4 (Figure 9.5) leaves the indecision zone to the right, so we can accept the lot. No further samples are needed.

Table 9.1. Sequential sampling plan for code letter L and acceptable quality level of 0.65%.

Sample No.	Sample Size	Cumulative Sample Size	Acceptance Number, Ac	Rejection Number, Re
1	50	50	#	3
2	50	100	0	3
3	50	150	1	4
4	50	200	2	5
5	50	250	3	6
6	50	300	4	6
7	50	350	6	7

Note: The symbol # means that an accept decision is not permitted for this sample size.

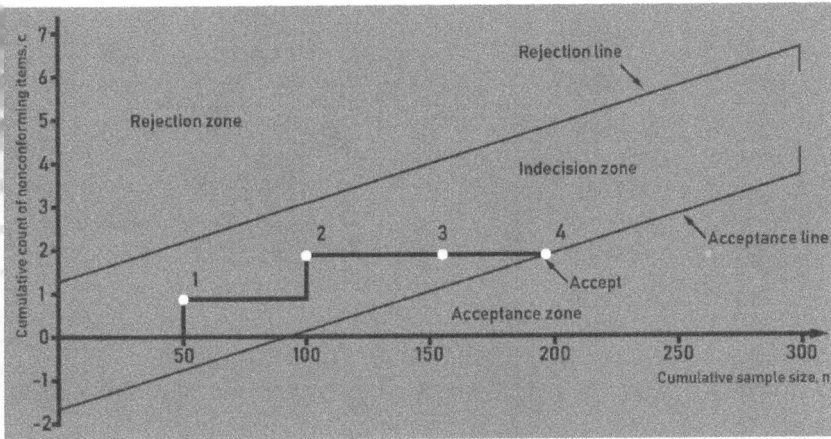

Figure 9.5. Sequential sampling plan.

Theoretically, the process of taking sequential samples may last forever without a chance of judging the lot. This is why the process is usually stopped when the accumulated sample size reaches the sample size of a single sampling plan and a decision is taken on this basis.

9.9 Comparative Analysis of Sampling Plans

Table 9.2 shows a comparison between the most popular AS plans in respect to some important features and facilitates the selection of the best one. For example, it could be seen that the single sampling plan has the lowest administrative costs while the sequential plan features the minimum number of inspected items per lot.

9.10 Designing a Standard Sampling Plan

General procedure:

(1) Choose/Agree upon the acceptable quality level (AQL)
    The AQL is the upper limit of the percentage of defective items (or the maximum number of defects per 100 items) that is acceptable as being satisfactory in terms of production process average.
    The AQL (typically between 0.01 and 10% defective items) is usually agreed with the customer in advance. AQL is the primary issue in the

Table 9.2. Comparison between the most popular AS plans.

Feature	Single Sampling	Double Sampling	Multiple Sampling	Sequential
Acceptability to producer	Psychologically poor to give only one chance of passing the lot	Psychologically adequate	Psychologically open to criticisms as being indecisive	Psychologically open to criticisms as being more indecisive than multiple sampling
Number of pieces inspected per lot	Generally greatest	Usually (but not always) 10–50% less than single sampling	Generally (but not always) less than double sampling by amounts of the order of 30%	Minimum over all attributes plans
Administrative cost in training personnel, records, drawing and identifying samples, etc.	Lowest	Greater than single sampling	Greater	Greatest
Information about prevailing level of quality in each lot	Most	Less than single sampling	Less than double sampling	Least

standard. Different AQLs may be designated for different types of defects.

Note that AQL might be the maximum number of defects per 100 items (not percentage), which is why it may go over 100.

(2) The inspection level designates the relative amount of inspection. Choose the inspection level as follows:

- Inspection level I: when the inspection cost is comparatively high.
- Inspection level II: ordinary cases.
- Inspection level III: when the inspection costs are low.
- Special inspection levels S1–S4: when the cost of destructive test is high.

The inspection level effects the risks and inspection costs. For example, on having a relatively smaller sample size, the special inspection levels (S) will reduce the cost of the inspection but will increase the risk for misjudgement.

(3) According to the lot size and the chosen inspection level, determine the code letter using Table 9.3.
(4) Determine the appropriate type of sampling plan, e.g. single or double.
(5) Using the relevant standard plan tables (BS 6001-1:1999, ISO 2859-1:1999), find the appropriate sample size and Ac, Re numbers needed to build and implement the plan.

For example, using Table 9.3 and the relevant table from Appendix B (single or double sampling plan), for normal inspection level with

Table 9.3. Sample size code letters (ISO 2859–1999).

Lot Size	Special Inspection Levels				General Inspection Levels		
2–8	A	A	A	A	A	A	B
9–15	A	A	A	A	A	B	C
16–25	A	A	B	B	B	C	D
...	...	...	...	...	...	...	...
1,201–3,200	C	D	E	G	H	K	L
3,201–1,0000	C	D	F	G	J	L	M
10,001–35,000	C	D	F	H	K	M	N
35,001–150,000	D	E	G	J	L	N	P
150,001–500,000	D	E	G	J	M	P	Q
500,000 and over	D	E	H	K	N	Q	R

AQL = 1%, we can find that for a lot size 5000 (code letter L), the recommended sample size is 200 and Ac = 5, Re = 6.

9.11 Severity of Inspection

In addition to the code letters, the standards for sampling plans include tables for:

- A *normal inspection* plan — used when the process is considered to be operating at, or slightly better than, the AQL.
- A *tightened inspection* plan — using stricter acceptance criteria than those used in normal inspection. It is recommended when there is evidence for problems with the quality.
- A *reduced inspection* plan, which permits smaller sample sizes hence, it is cheaper than normal inspection.

Figure 9.6 shows an example (standard single sampling plan with AQL = 1%) for the relationship between the probability P_a that the sampling plan will accept the lot (also known as probability of acceptance P_a) and the percentage of defectives p (also known as fraction defectives). The three curves represent the following:

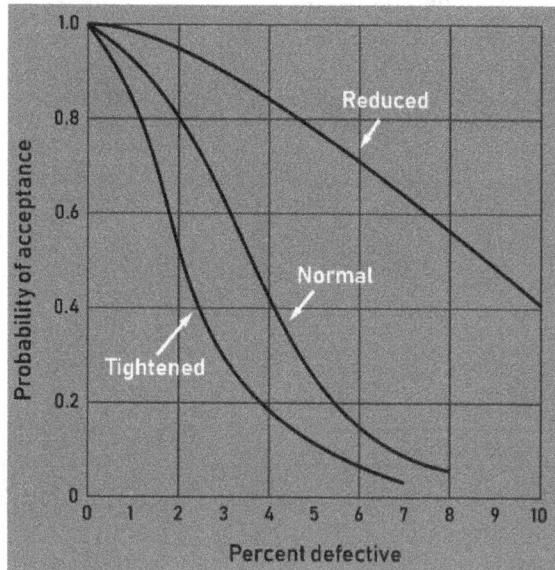

Figure 9.6. Operating curves.

- Normal inspection level: $n = 80$, Ac $= 2$, Re $= 3$.
- Tightened inspection level: $n = 80$, Ac $= 1$, Re $= 2$.
- Reduced inspection level: $n = 32$, Ac $= 1$, Re $= 2$.

It could be seen that the choice of severity of inspection depends on and affects the parameters of the plan. This is important because a lot with a given actual percentage of defectives, according to the applied severity of inspection, would have a different probability of acceptance and thus may be judged differently. For example, a lot with 4% defectives in it would have only about 20% (0.2) probability of acceptance if a tightened plan is used, about 40% with normal plan and over 80% with a reduced plan. The tightened plan has the lowest P_a because it has the strictest criteria (due to low acceptance and rejection numbers). The reduced plan would have the highest P_a because the judgement is based on an acceptance number Ac, which is relatively high, as it has the smallest sample size n.

9.11.1 *Evaluation of sampling plans*

Plans could be evaluated and compared using their operating characteristic (OC). Graphically, the OC is represented by the OC curve, e.g. see Figure 9.6.

9.11.1.1 *How to build OC curve*

To build an OC curve of a sampling plan, follow the step-by-step procedure as described as follows:

(1) Decide on n and c (e.g. $n = 80$, $c = 2$).
(2) Plot p (% fraction defective) on the abscissa scale, usually ranging from 0 to 10% (0–0.10), see Figure 9.7.
(3) Multiplying p by n, calculate expected number of defectives in the sample np for each fraction defective number in the batch (lot) $p = 1$, 2, 3..., 10%.
(4) From the cumulative Poisson distribution table (see Table B.9 in Appendix B) for each np and at given c (e.g. $c = 2$), obtain the probability of acceptance P_a value.
(5) Plot the fraction defective in the batch p versus the relevant probability of acceptance P_a.

p	np	Pa
0.00	0.00	1.0000
0.005	0.40	0.9921
0.01	0.80	0.9526
0.02	1.60	0.7834
0.03	2.40	0.5697
0.04	3.20	0.3799
0.05	4.00	0.2381
0.06	4.80	0.1425
0.07	5.60	0.0824
0.08	6.40	0.0463
0.09	7.20	0.0255
0.10	8.00	0.0138

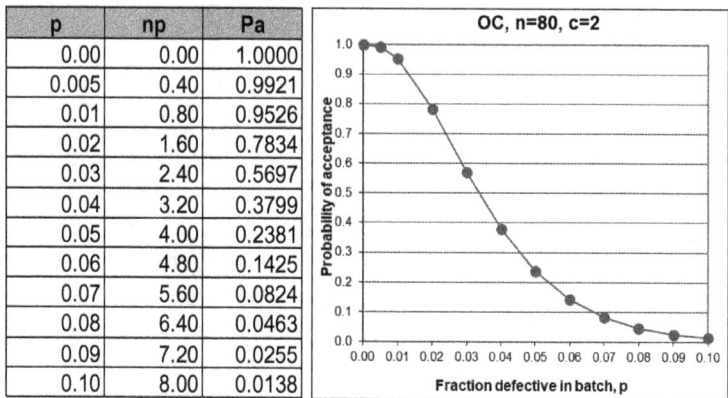

Figure 9.7.   Building OC for a plan with $n = 80$, $c = 2$.

9.11.1.2 *Interpretation of OC curves*

Typically, plans with steeper operating characteristic curves are considered more reliable because they have a better 'discriminating power', which is the ability of the plan to distinguish between good and bad lots. Obviously, from this point of view, the tightened and normal plans shown in Figure 9.6 are much more reliable than the reduced plan. However, the latter would be a lot cheaper to run as it is based on a smaller sample size but the risks of misjudgement would be higher.

9.11.2 *Switching rules*

Typically, an inspection starts at normal inspection level. Figure 9.8 illustrates the switching rules that apply for ongoing inspections.

9.12 Rectifying Inspection

When using AS, the naturally posed question is: What do we do with rejected lots? If the inspection is applied to incoming goods, then the lots which are not accepted by the sampling plan could just be returned to the supplier. However, if the AS is applied to outgoing goods, this is not an option. Another possibility would be to recycle the whole lot including the good parts, which may be rather wasteful. A third possibility known as

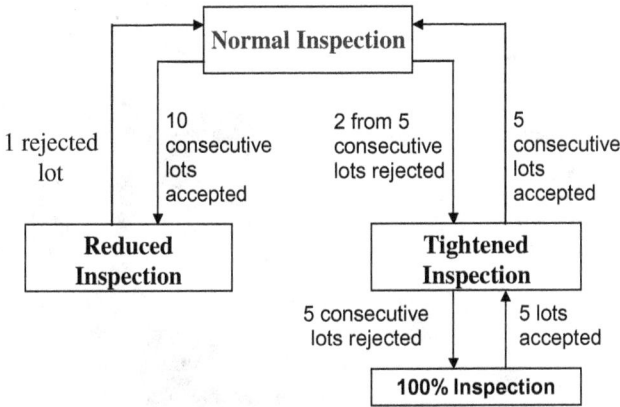

Figure 9.8. AS switching rules.

rectifying inspection is when the rejected lots are subject to a further 100% inspection and are cleared of all non-conforming items by replacing them with good items (Figure 9.9). The result, on the average, is that, after inspection, $100(1-P_a)$% lots are 100% conforming and $100P_a$% lots, which have been inspected by sampling alone and accepted, contain a percentage ($100p$) of non-conforming items (minus a few rejects replaced from the sample only rectifying it).

9.12.1 *Average outgoing quality and its limit (AOQL)*

The average outgoing quality (AOQ), in percent of non-conforming items that are passed on to a customer, will be approximately

$$AOQ = 100(P_a \times p)\%. \tag{9.1}$$

The above formula allows us to build the AOQ curve, which gives the average outgoing quality (left axis) as a function of the incoming quality, i.e. fraction defective in the batch (bottom axis/abscissa). Figure 9.10 shows an example of AOQ for a plan with sample size $n = 80$ and acceptance number $c = 2$. Note that the AOQ curve has a distinct maximum. The AOQL is the maximum (limit) of the worst possible defective or defect rate (AOQ) that should ever be expected for a given sampling plan. At first glance, this maximum (peak) seems hard to understand. Consider the

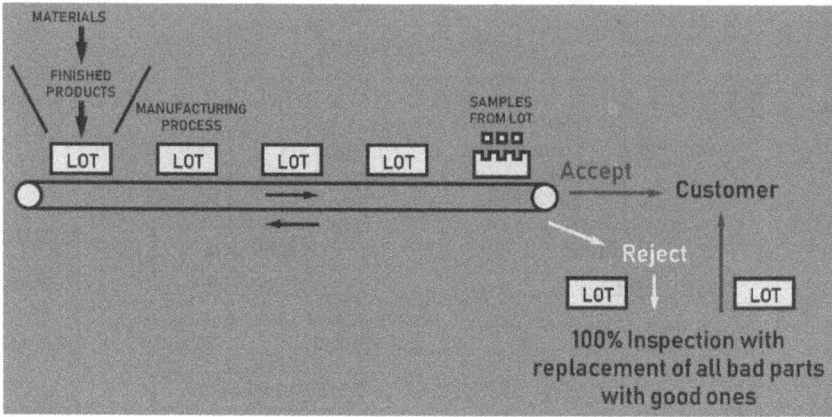

Figure 9.9.   Rectifying inspection.

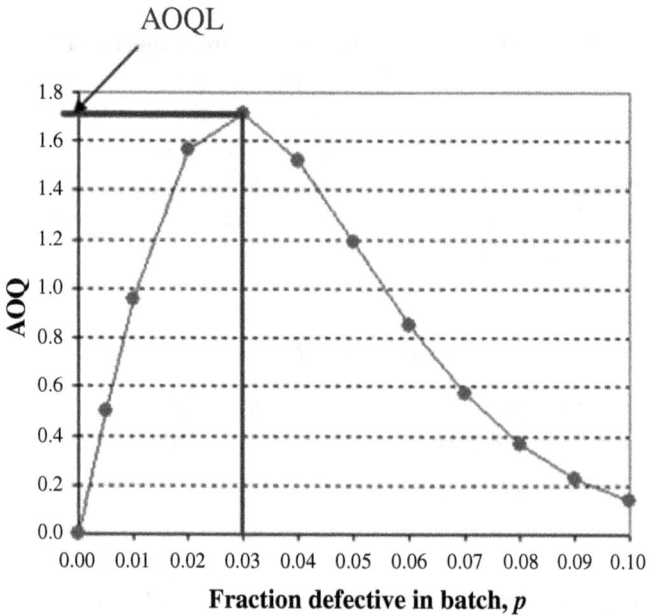

Figure 9.10.   AOQ and its maximum AOQL for $n = 80$, $c = 2$.

values along the abscissa. If the initial incoming quality is very good (fraction defective p_0 close to zero), the initial AOQ_0 values will naturally be low. With incoming quality deteriorating (increasing p), the AOQ will

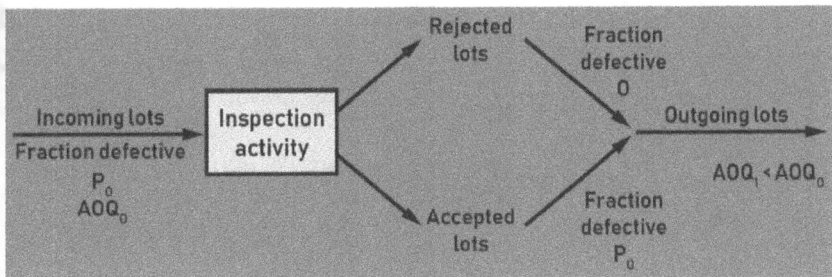

Figure 9.11.   Fraction defective in rectifying inspection.

increase too but only to a certain point/limit — AOQL, where the AS will start rejecting more lots which will be cleared off of the bad components ($p = 0$). Then, they are mixed with the accepted lots (see Figure 9.11) which will then start to reduce the AOQ to $AOQ_1 < AOQ_0$.

So, regardless of the incoming quality, the defective or defect rate going to the customer should be no greater than the AOQL over an extended period of time. Individual lots might be worse than the AOQL, but in the long run, the quality should not be worse than the AOQL.

This approximation is good if the lot size (N) is at least 10 times the sample size (n).

The rectifying inspection can be considered as an acceptance sampling add-on tool, which aims to improve the outgoing quality. However, it should be understood that it would also increase the cost of the product as there is a 100% inspection of some lots involved.

Activity 9.1:
Assuming normal inspection, obtain all required parameters and build block diagrams for standard single and double sampling plans, where AQL = 0.65% and lot size $N = 5000$. How would you judge the lot quality if, respectively, two defective items are found in the first sample and two defective items are found in the second sample both randomly taken from the lot? Compare these two sampling plans in general and in this particular case.

Chapter 10

Standardisation, Certification and Quality Management Systems

10.1 What is Standardisation?

Standardisation is the process of implementing and developing technical standards[1] based on the consensus of different parties, which may include companies, users, interest groups, standards organisations and governments. Standardisation can help maximise compatibility, interoperability, safety, repeatability or quality. The implementation of standards in industry and commerce became highly important with the beginning of the Industrial Revolution and the need for high-precision machine tools and interchangeable parts. Joseph Whitworth's screw thread measurements were adopted as the first (unofficial) national standard by companies around the country in 1841. It came to be known as the British Standard Whitworth, and it was widely adopted in other countries. A typical example for what the lack of standardisation may lead to is the Great Baltimore (United States) fire in 1904, which devastated the city because the firefighter's hose couplings were not standardised and they could not be connected or extended.

---

[1]A technical *standard* is an established norm or requirement for a repeatable technical task. It is usually a formal document that establishes uniform engineering or technical criteria, methods, processes, and practices. According to the application range, the standards can be classified as local, national and international. In addition, according to the power, the standards can be regarded as advisory (recommendations) or compulsory (regulations). Typical examples for compulsory standards are those related to health and safety.

10.2 International Organisation for Standardisation

The International Organisation for Standardisation (ISO) is a worldwide federation of national standards bodies (ISO member bodies). It is located in Switzerland and was established in 1947 to develop common international standards in many areas. Its members come from over 120 national standards bodies.

One of the duties of the ISO is to prepare and maintain the International Standards. The work of preparing International Standards is normally carried out through the ISO technical committees. Each member body interested in a subject, for which a technical committee has been established, has the right to be represented on that committee.

International Standards provide a reference framework, or a common technological language, between suppliers and their customers, which facilitates trade and the transfer of technology.

The International Standards can be used by internal and external parties, including certification bodies, to assess the organisation's abilities to meet customers, regulate and to see the organisation's own requirements. The ISO owns the registered trademarks for the ISO logo (Figure 10.1).

10.3 ISO 9000 Family of Standards

The term 'ISO 9000' refers to a set of quality management standards. Together, the ISO 9000 series of standards form a set of guidelines for implementing and operating an effective and efficient quality management system (QMS) in organisations of any type. Organisations use the standards to demonstrate the ability to consistently provide products and services that meet customer and regulatory requirements.

The ISO 9000 family includes four main quality standards presented in Table 10.1.

Figure 10.1.   ISO's logo.

Table 10.1. ISO 9000 family standards.

Name of Standard	Type of Standard	Description
ISO 9000:2015	Vocabulary	Contains QMS fundamentals and vocabulary
ISO 9001:2015	Requirement standard	Contains QMS requirements
ISO 9004:2009	Guidelines	Contains QMS guidelines for performance and improvement
ISO 9004:2011	Guidelines	Guidelines on internal and external QMS audits

Other standards related to QMSs include the ISO 14000 series (environmental management systems), ISO 13485 (QMSs for medical devices), ISO 19011 (auditing management systems) and ISO/TS 16949 (QMSs for automotive-related products).

10.3.1 *Main topics of the ISO 9001 standard*

ISO 9001 provides a comprehensive approach to documenting and reviewing the structure, responsibilities and procedures required to achieve effective quality management in an organisation. Specific sections of the standard contain information on many topics, such as:

- terms and definitions;
- requirements for a quality management system, including documented information, planning and determining process interactions;
- responsibilities of management;
- management of resources, including human resources and an organisation's work environment;
- product realisation, including the steps from design to delivery;
- measurement, analysis and improvement of the QMS through activities such as internal audits and corrective and preventive actions.

10.3.2 *Documented (written) procedures required by ISO 9001*

1. Control of documents (e.g. who can issue, sign and change documents).
2. Control of records (see the following section).
3. Internal audits: responsibilities, planning, conducting, reporting results and maintaining records.

4. Control of non-conforming products (what to do with non-conforming products).
5. Corrective and preventive actions.

10.3.3 *Records required by ISO 9001*

ISO 9001 requires the following records to be established and maintained:

- Management review (based on results of audits, customer feedback, process performance, planned changes and recommendations for improvements).
- Education, training, skills and experience.
- Review of requirements related to the product.
- Product requirements.
- Design and development (review, verification, validation, changes).
- Purchasing (including supplier evaluation).
- Validation of processes for production and services.
- Customer property.
- Control of measuring devices.
- Internal audit results.
- Monitoring and measuring of product.
- Nature of non-conformities.
- Corrective action results.
- Preventive action results.

10.3.4 *Changes to ISO 9001*

Since 1994, the ISO 9000 standards have been updated a number of times. The latest changes introduced in the 2015 ISO 9001 revision are intended to ensure that ISO 9001 continues to adapt to the changing environments in which organisations operate. Some of the key updates in ISO 9001:2015 include:

- the introduction of new terminology;
- a restructuring of some of the information;
- an emphasis on risk-based thinking to enhance the application of the process approach;
- improved applicability for services;
- increased leadership requirements.

10.4 Quality Management System

QMS is a part of the organisation's management system that focuses on the achievements related to *quality objectives*. The quality objectives complement other objectives of the organisation, such as profitability, growth, funding, environment protection, health and safety.

The quality objectives are declared in the *quality policy*, which is a document stating the overall intentions and directions of the organisation in respect to the quality expressed by the top management. The top management is a person or a group of people who direct and control the organisation at the highest level.

10.4.1 *Element of QMS*

The main elements of QMS are as follows:

1. Quality documents (quality policy, objectives and quality manual).
2. Organisational structure and responsibilities.
3. Data management.
4. Processes (including purchasing).
5. Product quality leading to customer satisfaction.
6. Continuous improvement including corrective and preventive action.
7. Quality instruments.
8. Document control.

A QMS can be product based or process based. ISO 9000 recommends process-based QMS. The main advantage of the process-based approach is that it provides control over linkage between the individual processes and their interaction. The model given in Figure 10.2 is based on plan-do-check-act methodology (Deming).

The QMS should be designed, operated and improved in accordance with the seven *quality management principles* which are defined in ISO 9000:2015, Quality management systems — Fundamentals and vocabulary. They are as follows:

- Customer focus.
- Leadership commitment to quality.
- Engagement of people.

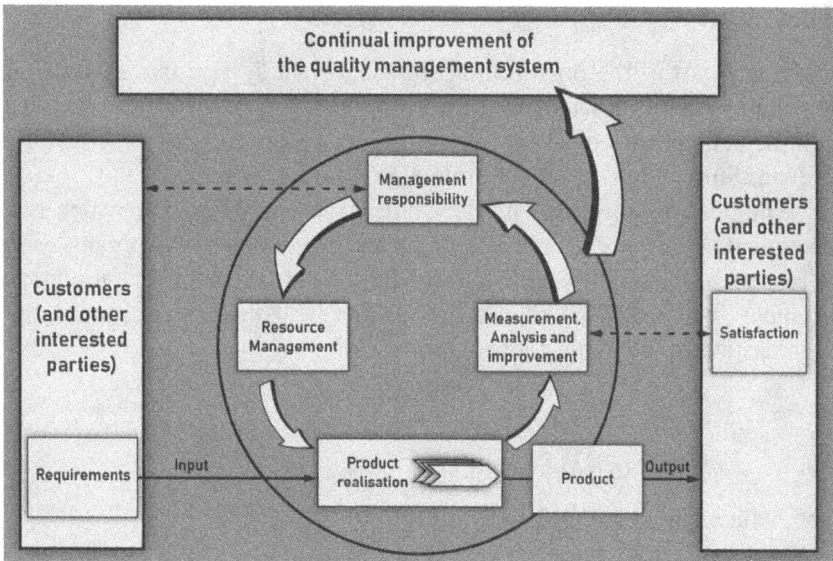

Figure 10.2.   Model of a process-based QMS.

- Process approach.
- Continuous improvement.
- Evidence-based decision making.
- Relationship management.

10.4.2 *QMS documentation*

QMS documentation is comprised of:

(1) *quality policy* containing the *quality objectives* and statements;
(2) *quality manual*, which is the main quality document, usually including the business scope, procedures and interactions between processes;
(3) *documented procedures and records* required by the ISO 9001 standard;
(4) *documents*, determined by the organisation to be necessary to ensure the effective planning, operation and control of its processes.

The term 'documented procedure', which appears within this International Standard, means that the procedure is established, documented, implemented and maintained. A single document may address the requirements for one or more procedures. A requirement for a documented procedure may be covered by more than one document.

The extent of the QMS documentation can differ from one organisation to another due to:

(a) the size of the organisation and the type of activities;
(b) the complexity of processes and their interactions;
(c) the competence of personnel.

The documentation can be in any form or type of medium.

10.5 Quality Audit

The term 'audit' is usually referred to as an act of a systematic formal evaluation and assessment of the accounts and/or activities carried out by a company aiming to ascertain the conformance of the matter. This can be done by self (first-party) assessment of an assessor(s), which is part of the company.

Conformance to one of the quality system standards may be specified between two parties as a contractual requirement. In this case, conformance to the standard may be assessed by the customer placing the contract

Figure 10.3. Types of quality audits.

(second-party assessment), see Figure 10.3. In many cases, when following such individual contract approach, an organisation is audited multiple times for essentially the same requirements. Alternatively, an independent assessment of a company's QMS can be carried out by an approved body (third-party assessment).

10.6 Quality Management Systems and Certification

10.6.1 *Quality system certification*

Certification means that a third-party organisation, which is called a 'certification body', conducts a formal audit of a supplier organisation to assess conformance to the appropriate quality system standard (ISO 9000). ISO 9001 is the only standard in the ISO 9000 series to which organisations can certify. Achieving ISO 9001 certification means that an organisation has demonstrated that it:

- follows the guidelines of the ISO 9001 standard;
- fulfils its own requirements;
- meets customer requirements and statutory and regulatory requirements;
- maintains documentation.

When the supplier organisation is judged to be in complete conformance, the third party issues a certificate to the supplying organisation and registers the organisation's quality system in a publicly available register, which is followed by a certificate. The development of QMS certification/registration is a mean to reduce the redundant and non-value-adding effort of these multiple audits.

Benefits of having an approved/certified QMS:

(a) Customers are satisfied because goods and services are always produced according to their requirements.
(b) It reduces operating costs as waste is eliminated and efficiency is increased as a result of eliminating non-conformance.
(c) It improves competitiveness and profitability as operating costs are reduced.
(d) It avoids duplication of customer assessments.
(e) It provides evidence of a responsible attitude to quality and product liability requirements.

10.7 Criticisms of QMS Registration

Some of the criticisms of the QMS registration are related to the cost, time and extra paperwork involved in the process. Most important costs incurred by the registration are as follows:

- Cost of an application fee. This varies depending on the size of the company and complexity of the business.
- Cost associated with conducting the required stage one and stage two audits for ISO 9001 certification.
- Hourly and per-day rates charged for off-site and on-site audit activities.
- Administrative fees if any.
- Travel and subsistence costs (minimum and maximum daily charges).
- Frequency and cost for surveillance audits to maintain certification.
- Cost for QMS recertification.

10.8 Process of Achieving ISO 9000 Certification

Each company, big or small, has some QMS. However, to get this QMS certified, a company should go through a registration process. Usually, the QMS certification process runs through several stages, as presented in Table 10.2, which gives some explanation as to what is happening during each stage. Depending on the status of the existing (pre-registered) QMS and the nature of the business, the preparation process can be fairly straightforward or it may be longer and a bit more complicated. Please note that the pre-certification audit stage (item 6 in Table 10.2) is optional, aiming to prevent rushed formal audits, which can be costly. Usually, the formal audit takes two to five working days on-site for completion, depending on the complexity of the business.

At the end of the process, the registrar will suggest a time (usually two to five years) when the next (re-certification) formal audit is needed to maintain the registration and when this should take place.

10.9 Accreditation

Registrars audit and certify organisations who wish to become ISO 9000 registered or certified.

Table 10.2.   Stages of the QMS certification process.

Stage	Name	What Happens?
1	Strategic Planning	Management: • shows commitment; • forms a project team (including a quality manager); • establishes a timeline; • selects a registrar (certification body).
2	Process Identification	• Identifies the processes needed. • Determines the sequence and the interaction between the processes.
3	Gap Analysis	• Evaluation of the existing QMS against the ISO (e.g. 9001) standard requirements. • Plan of *corrective actions, including targets and measures* to be monitored for effective implementation of a proper QMS.
4	Corrective Actions	• Corrective actions take place as identified in the previous stage. • *Internal audits* will be planned. (Trained internal auditors are required.)
5	Implementation of QMS	• Implementation of *document/form* structure and control system (including responsibilities). • Implementation of record control system. • Conducting *internal audits* (minimum three months are recommended prior to the formal audit).
6	Pre-Certification Audit	• Auditor ensures that the organisation is ready to be audited. A date for the formal audit will be suggested. All major non-conformities will be reported.
7	Certification Audit	• Documentation and QMS review. • The registrar will/will not recommend the organisation for registration reporting what (if any) non-conformities have been found.
8	ISO 9000 Registration	• If there is evidence that all major non-conformities have been addressed, an ISO 9000 certificate will be issued.

Accreditation bodies, on the other hand, evaluate and accredit registrars. In effect, accreditation bodies audit the auditors. Accreditation bodies certify that the registrars are competent and authorised to issue ISO 9000 certificates in specified business sectors.

So, the *accreditation* is a process aiming to assure the competence and the objectivity of accredited bodies, which are authorised to issue certificates. The accreditation has been set up world-wide. Accreditation bodies audit the registrars for conformity to standard international guides for the operation of certification bodies.

10.9.1 *National accreditation*

National accreditation bodies accredit organisations to perform formal audits and issue certificates [3]. Usually, the government licenses the national accreditation bodies.

For example, The United Kingdom (UK) national accreditation body, the UK Accreditation Service (UKAS), is evaluated internationally as working to the agreed international standards for the operation of accreditation bodies. The logo of UKAS is shown on Figure 10.4.

Certification bodies and laboratories accredited by UKAS are assessed to the appropriate and agreed international standards.

UKAS is recognised by the UK Government as the national body responsible for providing national accreditation of certification bodies (NACCB) and national accreditation of measurement and sampling (NAMAS).

The UK Government has licensed UKAS to use and sublicense to its accredited organisations the right to use the NACCB and the NAMAS

Figure 10.4.   UKAS logo.

logos, which incorporate the Royal Crown. For example, metrology labs accredited by NAMAS are authorised to check, calibrate and issue certificates for a certain type of measuring instruments. In this case, the accreditation confirms that these laboratories have the required competence and facilities/equipment and follow the correct procedures to certify measuring instruments.

Activity 10.1:
In the context of technical standardisation, comment on the advantages and disadvantages of the interchangeability as a principle used in engineering design. Give examples for full and limited interchangeability.

Activity 10.2:
Explain the difference between being certified and being accredited.

Activity 10.3:
What is the difference between being certified and being registered?

Appendices

Appendix A

Answers

Activity 1.1:
(1) Validation
(2) Test
(3) None of the suggested
(4) Inspection
(5) Verification

Activity 2.1:
(a)

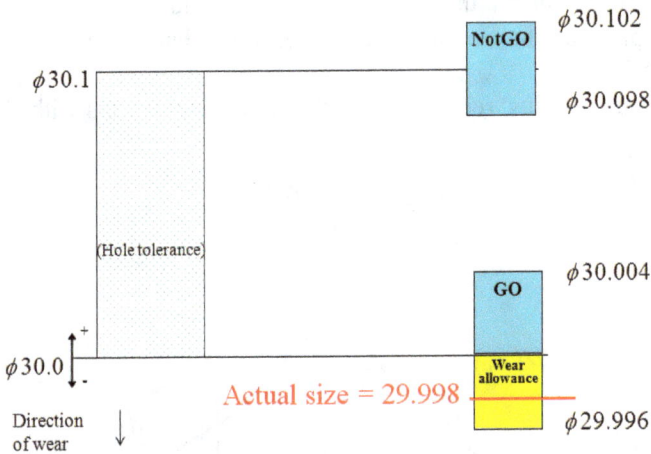

(b) Is the diameter good? No, as it out of the tolerance $\varnothing 30^{+0.1}_{0}$, so should be a fail;

The actual hole diameter is smaller than the NOT-GO end, so it will not fit in — that is a pass, so far;

The hole smaller than the GO end of a new gauge will not fit in — fail as it is too small a hole. However, a worn gauge would be below 29.998 mm and the GO end will fit in and will pass it, which would be wrong!

Activity 3.1:
Tolerance $\varnothing 50 \pm 0.15$ mm could be represented by limits => upper $\varnothing 50.15$ mm and lower $\varnothing 49.85$ mm.

We need to do the Z transformation for both limits as targets:

$$Z_1 = \frac{T - \bar{X}}{\sigma} = \frac{49.85 - 50.05}{0.1} = -2.$$

To work out the probability represented by the area under the curve, from the standard (unit) normal cumulative distribution function, find $\Phi(Z_1) = \Phi(-2) = 0.02275 = 2.275\%$.

So, $0.02275 = 2.275\%$ is the fraction area (the entire area under the curve is 100%) under the curve on the left-hand side of the target Z_1, which represents the 'minus' rejects, i.e. those that fall below the lower tolerance limit.

To find the 'plus' rejects, repeat the same calculations with Z_2:

$$Z_2 = \frac{50.15 - 50.05}{0.1} = 1.$$

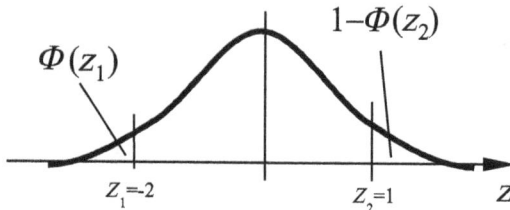

From the standard (unit) normal cumulative distribution function, find $\Phi(Z_2) = \Phi(1) = 0.84134 = 84.134\%$, which is the area under the curve again on the left-hand side of the target Z_2. To find the 'plus' rejects, i.e. those that fall above the upper tolerance limit, we need to calculate: $1 - \Phi(Z_2) = 1 - 0.84134 = 0.15866 = 15.866\%$.

Finally, the total probability of rejects (plus and minus) would be:

$$0.02275 + 0.15866 = 0.18141 = 18.141\%.$$

In a batch of 1 million, it would be expected to have:

$0.02275 \times 1,000,000 = 22,750$ minus rejects;
$0.15866 \times 1,000,000 = 158,660$ plus rejects;
total of 181,410 rejects.

Activity 3.2:
Unlike the method explained in 3.1, judging just by the number of defectives found in a sample would be a very rough estimate as it does not take into account the nature of the distribution, location of the average of the output in respect to (the middle of) the tolerance limits, standard deviation, etc.

For example, when applying this simplified method, if we take a sample of only five, it is quite likely that there will be zero rejects within this five. Then, an assumption that we would have zero defects in the whole batch would be really unsustainable.

Activity 4.3:
For dataset C:

Average $= 11.36$;
SD $= 6.23$;
ABS(Average $- 30) = 18.64$;
$3SD = 18.70 > 18.64$, hence according to the 3SD criteria, 30 is not an outlier;
t (for $n = 11$) $= 2.00$;
$tSD = 12.47 < 18.64$, hence according to tSD criteria, 30 is an outlier.

Activity 4.4:
As the mean value includes all values from the dataset, it will be affected by the presence of an outlier.

As the median is equal to only the middle member of the dataset, no matter how large or small the extreme values (minimum or maximum, which are the potential outliers) are, they will not have any effect on the value of the median at all.

Activity 4.5:

The first part of the procedure is to check data for outliers, which may take several iterations:

Original Set									
18.0	16.2	15.3	12.0	12.5	15.5	13.8	17.6	14.8	15.8
18.7	17.6	14.2	15.8	13.4	17.0	14.6	17.2	17.4	17.0
19.2	15.2	16.0	19.4	14.0	17.5	16.3	17.0	22.5	15.4
16.8	14.3	14.6	130	16.4	11.5	14.8	12.9	16.5	18.2
15.5	18.2	15.9	18.4	13.5	15.4	12.5	16.3	18.2	14.8
16.1	15.5	17.3	16.6	18.1	14.6	14.9	18.9	12.8	15.1
	Min	11.5							
	Max	130.0							
	Mean	17.86							
	SD	14.87	3SD	44.596					
	Outliers	−26.74	62.45		Outlier	130			

Set-1									
18.0	16.2	15.3	12.0	12.5	15.5	13.8	17.6	14.8	15.8
18.7	17.6	14.2	15.8	13.4	17.0	14.6	17.2	17.4	17.0
19.2	15.2	16.0	19.4	14.0	17.5	16.3	17.0	22.5	15.4
16.8	14.3	14.6		16.4	11.5	14.8	12.9	16.5	18.2
15.5	18.2	15.9	18.4	13.5	15.4	12.5	16.3	18.2	14.8
16.1	15.5	17.3	16.6	18.1	14.6	14.9	18.9	12.8	15.1
	Min	11.5							
	Max	22.5							
	Mean	15.96							
	SD	2.07	3SD	6.21					
	Outliers	9.74	22.17		Outlier	22.5			

(Continued)

Set-2									
15.5	14.3	14.2	12.0	12.5	11.5	12.5	12.9	12.8	14.8
16.1	15.2	14.6	15.8	13.4	14.6	13.8	16.3	14.8	15.1
16.8	15.5	15.3	16.6	13.5	15.4	14.6	17.0		15.4
18.0	16.2	15.9	18.4	14.0	15.5	14.8	17.2	16.5	15.8
18.7	17.6	16.0	19.4	16.4	17.0	14.9	17.6	17.4	17.0
19.2	18.2	17.3		18.1	17.5	16.3	18.9	18.2	18.2
	Min	11.5							
	Max	19.4							
	Mean	15.84							
	SD	1.90	3SD	5.69					
	Outliers	**10.15**	**21.54**		**No More Outlier**				

After removing all (two) outliers, the important statistical characteristics can be calculated as follows:

Range, R = max – min = 7.9.
Number of data in the dataset after removing the outliers = 58.
Number of classes for the histogram, K = 6.64, rounded up to 7.
Width of the class $d = R/K = 1.19$.

On this basis, the data can be classified into the following classes:

Lower Limit	Upper Limit	Class	Frequency
11.50	12.69	11.50–12.69	4
12.69	13.88	12.69–13.88	5
13.88	15.07	13.88–15.07	10
15.07	16.26	15.07–16.26	14
16.26	17.45	16.26–17.45	12
17.45	18.64	17.45–18.64	9
18.64	19.83	18.64–19.82	4
			58

Using the table below, the χ^2 table can be completed as follows:

Class j	Class Interval $Lj \leq Xj < Uj$	Interval Middle Xmj	$Z_j = (X_{mj} - \bar{X})/S$	$f(Z_j)$	faj	$ftj = f(Zj)*dl$ $S*N$	faj'	ftj'	$(faj' - ftj')^\wedge 2/$ ftj'
1	11.50–12.69	12.09	−1.98	0.056668	4	2.06			
2	12.69–13.88	13.28	−1.35	0.160654	5	5.84	9	7.90	0.15
3	13.88–15.07	14.47	−0.72	0.307352	10	11.17	10	11.17	0.12
4	15.07–16.26	15.66	−0.10	0.397118	14	14.43	14	14.43	0.01
5	16.26–17.45	16.85	0.53	0.346531	12	12.59	12	12.59	0.03
6	17.45–18.64	18.04	1.16	0.204223	9	7.42	13	10.37	0.66
7	18.64–19.82	19.23	1.78	0.081284	4	2.95			
					58	56.46			0.98

Note that the top and bottom classes have been joined with the neighbouring classes.

Now, the histogram can be constructed:

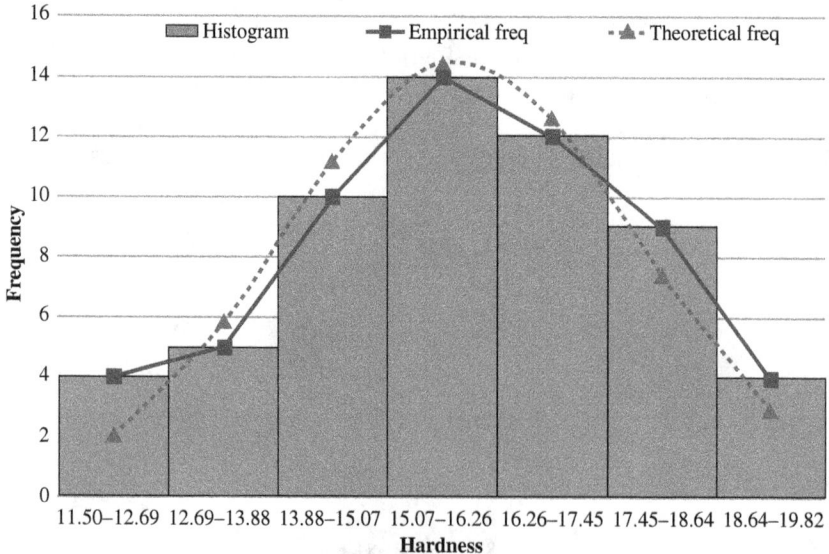

The tabulated $\chi^2_{\alpha,v}$ values can be obtained as follows:
$\chi^2_{0.05,2} = 5.991,$

where

$\alpha = 5\%$ or 0.05 and

$v = m - r - 1 = 5 - 2 - 1 = 2.$

As the calculated $\chi^2 = 5.991 < \chi^2_{0.05,2} = 5.991$, the normal distribution hypothesis can be accepted.

Activity 5.1:

In Figure 5.1, the tolerance T is wider than the process variation V, so $C_p > 1$. As there is no significant shift between the average and the middle of the tolerance, it is expected that k is close to 0, so $C_{pk} > 1$ too.

This means that the process is capable and well centred.

In Figure 5.2, the tolerance T is smaller than the process variation V, so $C_p < 1$. As $C_{pk} <= C_p$, hence $C_{pk} < 1$ too. This is an indication that this process capability is unsatisfactory. This means that regardless of the process centring, there will be a significant amount of scrap.

In Figure 5.3, the tolerance T seems wider than the process variation V, so $C_p > 1$. However, as there is a significant shift between the average and the middle of the tolerance, which will make k bigger, so it is expected that $C_{pk} < 1$ too. In this case, the process is capable, but because of the poor centring (shifted setting), there will be some scrap produced.

Activity 5.2:

$T = 4.$

$T_m = 32.$

$C_p = T/6SD = 4/3 = 1.33 =>$ very good potential process capability.

$C_{pk} = Cp(1 - k)$, $k = ABS(T_m - \bar{X})/0.5T = ABS(32-30)/2 = 1$, so $C_{pk} = 0.$

Ultimately, despite the good potential, the process performance will be poor because of the wrong setting.

Activity 5.3:

From the formula $C_{pk} = C_p(1 - k)$, it is obvious that C_p can be equal to C_{pk} if only $k = 0$. This would be only possible in the ideal case in which the process is dead well centred and the average of the output coincides with the middle of the tolerance.

Activity 5.4:

Again, considering the formula $C_{pk} = C_p (1 - k)$, it is obvious that $C_p = T/V$ would always be positive because it is a ratio of two positive variables.

So, C_{pk} can be negative only if $(1 - k)$ is negative. This would be only possible if $k > 1$. Considering the formula $k = \text{ABS}(T_m - \bar{X})/0.5T$, it is obvious that in this ratio, if $(T_m - \bar{X}) > 0.5T$, then $k > 1$. This means that if in the extreme case in which the shift between the average and the middle of the tolerance T_m is so big that it is larger than half of the tolerance, then $k > 1$, which will make C_{pk} negative.

Activity 6.2:

(a) Observing all standard out-of-control conditions, we can identify that in the R-chart area in sample 4 there is a point falling outside of the control limits. As the R-chart area represents the spread of the data, we could assume that there is some abnormality in that sample. If we study the data in sample 4, we would notice one value (40) which is much larger than the rest of the data. So, we could expect that 40 is an outlier, which is causing the spike in the R-chart.

(b) If we experiment with the data in Table 6.4, we will notice that both CCs, $\bar{X}/R$ and $\bar{X}/S$ will detect the outlier in sample 4, so they both have a very good sensitivity to such abnormalities.

(c) The control limits are deigned to work as warning levels, which are supposed to flag up any abnormalities before it is too late, i.e. before significant amount of scrap has been produced. For this reason, the control limits are supposed to be narrower and located inside the specification limits.

(d) Yes, it is possible if the average of the preliminary 20 samples used to establish the centre line CL and the control limits are shifted/deviate from the middle of the tolerance T_m. This would then cause a process that seems to be in control with no points outside of the control limits to have parts made out of the tolerance.

Activity 6.3:

(a) The *pn*-chart shown in Figure 6.10 clearly indicates two abnormalities in samples 20 and 21. The process is out of control.

(b) No. In variable CCs, a run of seven to eight points all above/below the centre line, which is a non-random patter, would be an indication of a shift of the process setting, whereas in attribute CCs, such abnormality is unlikely to be related to process settings.

(c) Looking at the formulas for calculating the control limits for both CCs, we could notice that subgroup size n is in the denominator. This means that the greater the subgroup size n, the narrower the control limits range and vice versa.

Activity 7.1:
- Testing TV sets: defective cause check sheet.
- Testing material hardness: production process distribution check sheet.
- Testing a large casting: defective location check sheet.
- Car MOT: checklist.

Activity 7.2:
Typical fishbone diagram:

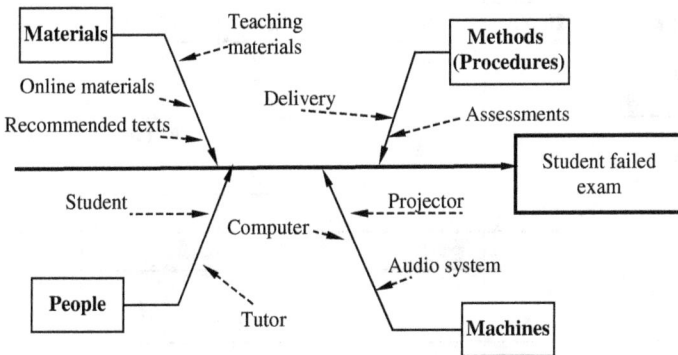

Activity 7.3:
(1)

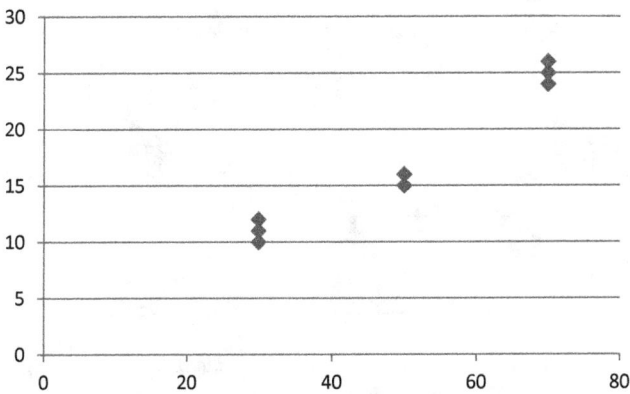

Scatter diagram

(2) Suggested linear model: $Y = aX + b$,

X	Y	Xi-AverX	(Xi-AverX)^2	Yi-AverY	(Xi-AverX) . (Yi-AverY)	(Yi-AverY)^2
30	10	−20.00	400.00	−7.11	142.22	50.57
30	12	−20.00	400.00	−5.11	102.22	26.12
30	11	−20.00	400.00	−6.11	122.22	37.35
50	15	0.00	0.00	−2.11	0.00	4.46
50	16	0.00	0.00	−1.11	0.00	1.23
50	15	0.00	0.00	−2.11	0.00	4.46
70	25	20.00	400.00	7.89	157.78	62.23
70	26	20.00	400.00	8.89	177.78	79.01
70	24	20.00	400.00	6.89	137.78	47.46
		Sum =	2400.00		840.00	312.89
Aver X	Aver Y					
50.00	17.11				r =	0.969
					r² =	0.9396
a =	Σ(Xi-AverX). (Yi-AverY)		=	0.35		
	Σ(Xi-AverX)^2					
b =	AverY-a.AverX		=	−0.39		

Regression model: $Y = 0.35X - 0.3889$.

(3)

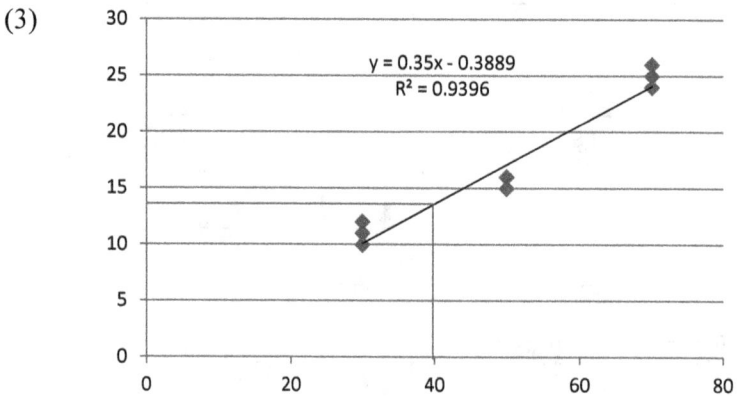

Fuel cost at 40 miles/hour:
$$Y = 0.35 \times 40 - 0.3889 = 13.61.$$

Activity 8.1:

Six-Sigma quality requires two conditions:

(1) $C_p => 2$.

(2) The average should be contained in the interval ± 1.5SD around the middle of the tolerance T_m.

The tolerance of 300 ± 6 could be represented by limits — upper 306 and lower 294. Tolerance $T = 12$, middle of the tolerance $T_m = 300$.

$C_p = T/6$SD $= 12/6 = 2$, so condition 1 is met.

In this case, the interval ± 1.5SD around T_m would be from $(300 - 1.5)$ $= 298.5$ to $(300 + 1.5) = 301.5$.

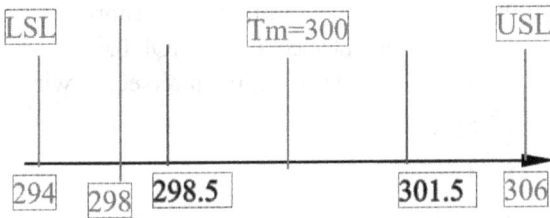

As it can be seen from the picture above, the average 298 falls outside the interval; this means the second condition has not been met. It seems that the problem is in the setting of the process, which drifted away of the middle of the tolerance T_m. The process should be stopped and setting adjusted. Then, another sample could be taken to check if Six-Sigma quality is achieved after resetting. However, even if it did, the process setting should be checked on a regular basis to avoid drifting off too much again.

Activity 9.1:

Single sampling plans:

1. Identify the code letter.
   From Table 9.3, for general inspection level II (ordinary case) and lot size between 3,000 and 10,000: L.
2. Then, from the table of single sampling plan (normal inspection), get sample size $n = 200$ and intersect L with AQL 0.65%: Ac = 3, Re = 4.

```
        ┌─────────┐
        │ N = 5000│
        └────┬────┘
             │
        ┌────┴────┐
        │ n = 200 │
        └────┬────┘
             │
          ◇──┴──◇
     ┌────┘     └────┐
┌──────────┐    ┌──────────┐
│z ≤ Ac = 3│    │z ≥ Re = 4│
└────┬─────┘    └────┬─────┘
     │               │
┌──────────┐  ┌─────────────┐
│Accept the│  │Do not accept│
│   lot    │  │   the lot   │
└──────────┘  └─────────────┘
```

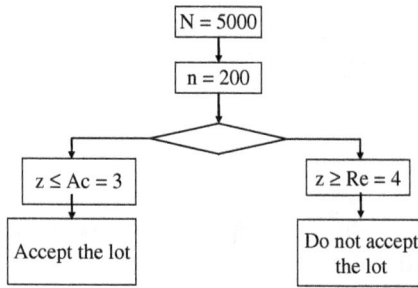

As $z = 3$: **accept the lot**.

Double sampling plan:

1. Identify the code letter. From Table 9.3, for general inspection level II (ordinary case) and lot size between 3,000 and 10,000: L.
2. Then, from the table of double sampling plan (normal inspection), get:
   – sample size $n_1 = 125$ and intersect L with AQL 0.65%: $Ac_1 = 1$, $Re_1 = 3$;
   – sample size $n_2 = 125$ (total of 250), intersect L with AQL 0.65%: $Ac_2 = 4$, $Re_2 = 5$.

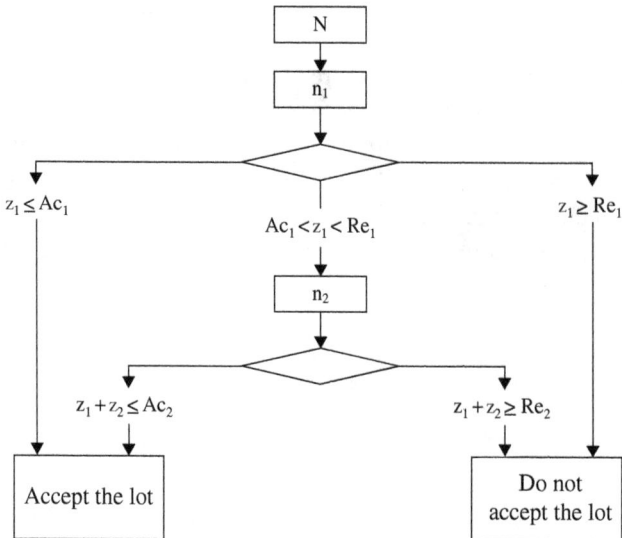

As $z_1 = 2$ which is between $Ac_1 = 1$ and $Re_1 = 3$, take a second sample of 125 (total of 250).

As $z_1 + z_2 = 4$ which is equal to $Ac_2 = 4$, **accept the lot**.

Single sample plan vs. double sampling plan comparison:

Generally, the double sampling plan is faster and cheaper than single sampling plan because the initial sample size is smaller. This applies to most cases, say 95%.

In this particular case (not typical) the double sampling plan was slower and dearer because it required a second sample, so the total of 250 is bigger than the sample size of the single sampling plan, but in return, this gave us a higher confidence in the decision we take as it is based on a larger sample.

Activity 10.1:
Interchangeability is a principle used in engineering design, which allows parts (units) manufactured independently to be assembled without further machining. For example, standard bolts, nuts, etc. fit in assemblies and work no matter where they have been produced as they are fully interchangeable. The main advantage of this principle is that standard parts/units can be replaced and the replacements will work the same way as the original parts. However, in order to achieve a full interchangeability, the parts must be designed according to a certain tolerance system. This may require the use of tight tolerances, which would make the production more expensive — the main disadvantage of the principle. A typical example for limited interchangeability is ball bearings. As the tolerance for the diameter of the balls is very tight, they are graded in groups and then assembled selectively, which makes the interchangeability limited within the group.

Activity 10.2:
Being certified means you are awarded a certificate. Being accredited means that you are authorised to issue certificates.

Activity 10.3:
It is the same thing. If you are registered it means you are certified.

Appendix B

Table B.1.   Standard normal cumulative distribution $\Phi(z)$, negative z.

z	0.00	0.01	0.02	0.03	0.04	0.05	0.06	0.07	0.08	0.09
-3.9	0.00005	0.00005	0.00004	0.00004	0.00004	0.00004	0.00004	0.00004	0.00003	0.00003
-3.8	0.00007	0.00007	0.00007	0.00006	0.00006	0.00006	0.00006	0.00005	0.00005	0.00005
-3.7	0.00011	0.00010	0.00010	0.00010	0.00009	0.00009	0.00008	0.00008	0.00008	0.00008
-3.6	0.00016	0.00015	0.00015	0.00014	0.00014	0.00013	0.00013	0.00012	0.00012	0.00011
-3.5	0.00023	0.00022	0.00022	0.00021	0.00020	0.00019	0.00019	0.00018	0.00017	0.00017
-3.4	0.00034	0.00032	0.00031	0.00030	0.00029	0.00028	0.00027	0.00026	0.00025	0.00024
-3.3	0.00048	0.00047	0.00045	0.00043	0.00042	0.00040	0.00039	0.00038	0.00036	0.00035
-3.2	0.00069	0.00066	0.00064	0.00062	0.00060	0.00058	0.00056	0.00054	0.00052	0.00050
-3.1	0.00097	0.00094	0.00090	0.00087	0.00084	0.00082	0.00079	0.00076	0.00074	0.00071
-3.0	0.00135	0.00131	0.00126	0.00122	0.00118	0.00114	0.00111	0.00107	0.00104	0.00100
-2.9	0.00187	0.00181	0.00175	0.00169	0.00164	0.00159	0.00154	0.00149	0.00144	0.00139
-2.8	0.00256	0.00248	0.00240	0.00233	0.00226	0.00219	0.00212	0.00205	0.00199	0.00193
-2.7	0.00347	0.00336	0.00326	0.00317	0.00307	0.00298	0.00289	0.00280	0.00272	0.00264
-2.6	0.00466	0.00453	0.00440	0.00427	0.00415	0.00402	0.00391	0.00379	0.00368	0.00357
-2.5	0.00621	0.00604	0.00587	0.00570	0.00554	0.00539	0.00523	0.00508	0.00494	0.00480
-2.4	0.00820	0.00798	0.00776	0.00755	0.00734	0.00714	0.00695	0.00676	0.00657	0.00639
-2.3	0.01072	0.01044	0.01017	0.00990	0.00964	0.00939	0.00914	0.00889	0.00866	0.00842
-2.2	0.01390	0.01355	0.01321	0.01287	0.01255	0.01222	0.01191	0.01160	0.01130	0.01101
-2.1	0.01786	0.01743	0.01700	0.01659	0.01618	0.01578	0.01539	0.01500	0.01463	0.01426
-2.0	0.02275	0.02222	0.02169	0.02118	0.02068	0.02018	0.01970	0.01923	0.01876	0.01831

Table B.1. (Continued)

z	0.00	0.01	0.02	0.03	0.04	0.05	0.06	0.07	0.08	0.09
-1.9	0.02872	0.02807	0.02743	0.02680	0.02619	0.02559	0.02500	0.02442	0.02385	0.02330
-1.8	0.03593	0.03515	0.03438	0.03362	0.03288	0.03216	0.03144	0.03074	0.03005	0.02938
-1.7	0.04457	0.04363	0.04272	0.04182	0.04093	0.04006	0.03920	0.03836	0.03754	0.03673
-1.6	0.05480	0.05370	0.05262	0.05155	0.05050	0.04947	0.04846	0.04746	0.04648	0.04551
-1.5	0.06681	0.06552	0.06426	0.06301	0.06178	0.06057	0.05938	0.05821	0.05705	0.05592
-1.4	0.08076	0.07927	0.07780	0.07636	0.07493	0.07353	0.07215	0.07078	0.06944	0.06811
-1.3	0.09680	0.09510	0.09342	0.09176	0.09012	0.08851	0.08691	0.08534	0.08379	0.08226
-1.2	0.11507	0.11314	0.11123	0.10935	0.10749	0.10565	0.10383	0.10204	0.10027	0.09853
-1.1	0.13567	0.13350	0.13136	0.12924	0.12714	0.12507	0.12302	0.12100	0.11900	0.11702
-1.0	0.15866	0.15625	0.15386	0.15151	0.14917	0.14686	0.14457	0.14231	0.14007	0.13786
-0.9	0.18406	0.18141	0.17879	0.17619	0.17361	0.17106	0.16853	0.16602	0.16354	0.16109
-0.8	0.21186	0.20897	0.20611	0.20327	0.20045	0.19766	0.19489	0.19215	0.18943	0.18673
-0.7	0.24196	0.23885	0.23576	0.23270	0.22965	0.22663	0.22363	0.22065	0.21770	0.21476
-0.6	0.27425	0.27093	0.26763	0.26435	0.26109	0.25785	0.25463	0.25143	0.24825	0.24510
-0.5	0.30854	0.30503	0.30153	0.29806	0.29460	0.29116	0.28774	0.28434	0.28096	0.27760
-0.4	0.34458	0.34090	0.33724	0.33360	0.32997	0.32636	0.32276	0.31918	0.31561	0.31207
-0.3	0.38209	0.37828	0.37448	0.37070	0.36693	0.36317	0.35942	0.35569	0.35197	0.34827
-0.2	0.42074	0.41683	0.41294	0.40905	0.40517	0.40129	0.39743	0.39358	0.38974	0.38591
-0.1	0.46017	0.45620	0.45224	0.44828	0.44433	0.44038	0.43644	0.43251	0.42858	0.42465
0.0	0.50000	0.49601	0.49202	0.48803	0.48405	0.48006	0.47608	0.47210	0.46812	0.46414

Quality Management Essentials

Table B.2. Standard normal cumulative distribution $\Phi(z)$, positive z.

z	0.00	0.01	0.02	0.03	0.04	0.05	0.06	0.07	0.08	0.09
0.0	0.50000	0.50399	0.50798	0.51197	0.51595	0.51994	0.52392	0.52790	0.53188	0.53586
0.1	0.53983	0.54380	0.54776	0.55172	0.55567	0.55962	0.56356	0.56749	0.57142	0.57535
0.2	0.57926	0.58317	0.58706	0.59095	0.59483	0.59871	0.60257	0.60642	0.61026	0.61409
0.3	0.61791	0.62172	0.62552	0.62930	0.63307	0.63683	0.64058	0.64431	0.64803	0.65173
0.4	0.65542	0.65910	0.66276	0.66640	0.67003	0.67364	0.67724	0.68082	0.68439	0.68793
0.5	0.69146	0.69497	0.69847	0.70194	0.70540	0.70884	0.71226	0.71566	0.71904	0.72240
0.6	0.72575	0.72907	0.73237	0.73565	0.73891	0.74215	0.74537	0.74857	0.75175	0.75490
0.7	0.75804	0.76115	0.76424	0.76730	0.77035	0.77337	0.77637	0.77935	0.78230	0.78524
0.8	0.78814	0.79103	0.79389	0.79673	0.79955	0.80234	0.80511	0.80785	0.81057	0.81327
0.9	0.81594	0.81859	0.82121	0.82381	0.82639	0.82894	0.83147	0.83398	0.83646	0.83891
1.0	0.84134	0.84375	0.84614	0.84849	0.85083	0.85314	0.85543	0.85769	0.85993	0.86214
1.1	0.86433	0.86650	0.86864	0.87076	0.87286	0.87493	0.87698	0.87900	0.88100	0.88298
1.2	0.88493	0.88686	0.88877	0.89065	0.89251	0.89435	0.89617	0.89796	0.89973	0.90147
1.3	0.90320	0.90490	0.90658	0.90824	0.90988	0.91149	0.91309	0.91466	0.91621	0.91774
1.4	0.91924	0.92073	0.92220	0.92364	0.92507	0.92647	0.92785	0.92922	0.93056	0.93189
1.5	0.93319	0.93448	0.93574	0.93699	0.93822	0.93943	0.94062	0.94179	0.94295	0.94408
1.6	0.94520	0.94630	0.94738	0.94845	0.94950	0.95053	0.95154	0.95254	0.95352	0.95449
1.7	0.95543	0.95637	0.95728	0.95818	0.95907	0.95994	0.96080	0.96164	0.96246	0.96327
1.8	0.96407	0.96485	0.96562	0.96638	0.96712	0.96784	0.96856	0.96926	0.96995	0.97062
1.9	0.97128	0.97193	0.97257	0.97320	0.97381	0.97441	0.97500	0.97558	0.97615	0.97670

Table B.2. (*Continued*)

z	0.00	0.01	0.02	0.03	0.04	0.05	0.06	0.07	0.08	0.09
2.0	0.97725	0.97778	0.97831	0.97882	0.97932	0.97982	0.98030	0.98077	0.98124	0.98169
2.1	0.98214	0.98257	0.98300	0.98341	0.98382	0.98422	0.98461	0.98500	0.98537	0.98574
2.2	0.98610	0.98645	0.98679	0.98713	0.98745	0.98778	0.98809	0.98840	0.98870	0.98899
2.3	0.98928	0.98956	0.98983	0.99010	0.99036	0.99061	0.99086	0.99111	0.99134	0.99158
2.4	0.99180	0.99202	0.99224	0.99245	0.99266	0.99286	0.99305	0.99324	0.99343	0.99361
2.5	0.99379	0.99396	0.99413	0.99430	0.99446	0.99461	0.99477	0.99492	0.99506	0.99520
2.6	0.99534	0.99547	0.99560	0.99573	0.99585	0.99598	0.99609	0.99621	0.99632	0.99643
2.7	0.99653	0.99664	0.99674	0.99683	0.99693	0.99702	0.99711	0.99720	0.99728	0.99736
2.8	0.99744	0.99752	0.99760	0.99767	0.99774	0.99781	0.99788	0.99795	0.99801	0.99807
2.9	0.99813	0.99819	0.99825	0.99831	0.99836	0.99841	0.99846	0.99851	0.99856	0.99861
3.0	0.99865	0.99869	0.99874	0.99878	0.99882	0.99886	0.99889	0.99893	0.99896	0.99900
3.1	0.99903	0.99906	0.99910	0.99913	0.99916	0.99918	0.99921	0.99924	0.99926	0.99929
3.2	0.99931	0.99934	0.99936	0.99938	0.99940	0.99942	0.99944	0.99946	0.99948	0.99950
3.3	0.99952	0.99953	0.99955	0.99957	0.99958	0.99960	0.99961	0.99962	0.99964	0.99965
3.4	0.99966	0.99968	0.99969	0.99970	0.99971	0.99972	0.99973	0.99974	0.99975	0.99976
3.5	0.99977	0.99978	0.99978	0.99979	0.99980	0.99981	0.99981	0.99982	0.99983	0.99983
3.6	0.99984	0.99985	0.99985	0.99986	0.99986	0.99987	0.99987	0.99988	0.99988	0.99989
3.7	0.99989	0.99990	0.99990	0.99990	0.99991	0.99991	0.99992	0.99992	0.99992	0.99992
3.8	0.99993	0.99993	0.99993	0.99994	0.99994	0.99994	0.99994	0.99995	0.99995	0.99995
3.9	0.99995	0.99995	0.99996	0.99996	0.99996	0.99996	0.99996	0.99996	0.99997	0.99997

Quality Management Essentials

Table B.3.   Chi-squared distribution, $\chi^2_{\alpha,\,\nu}$.

ν	α					
	0.5	**0.1**	**0.05**	**0.025**	**0.01**	**0.005**
1	0.455	2.706	3.841	5.024	6.635	7.879
2	1.386	4.605	5.991	7.378	9.210	10.597
3	2.366	6.251	7.815	9.348	11.345	12.838
4	3.357	7.779	9.488	11.143	13.277	14.860
5	4.351	9.236	11.070	12.833	15.086	16.750
6	5.348	10.645	12.592	14.449	16.812	18.548
7	6.346	12.017	14.067	16.013	18.475	20.278
8	7.344	13.362	15.507	17.535	20.090	21.955
9	8.343	14.684	16.919	19.023	21.666	23.589
10	9.342	15.987	18.307	20.483	23.209	25.188
11	10.341	17.275	19.675	21.920	24.725	26.757
12	11.340	18.549	21.026	23.337	26.217	28.300
13	12.340	19.812	22.362	24.736	27.688	29.819
14	13.339	21.064	23.685	26.119	29.141	31.319
15	14.339	22.307	24.996	27.488	30.578	32.801
16	15.338	23.542	26.296	28.845	32.000	34.267
17	16.338	24.769	27.587	30.191	33.409	35.718
18	17.338	25.989	28.869	31.526	34.805	37.156
19	18.338	27.204	30.144	32.852	36.191	38.582
20	19.337	28.412	31.410	34.170	37.566	39.997

Note: In Excel, use CHIINV(α, ν).

Table B.4.   Standard normal probability density function.

z +/-	0.00	0.01	0.02	0.03	0.04	0.05	0.06	0.07	0.08	0.09
0.0	0.3989	0.3989	0.3989	0.3988	0.3986	0.3984	0.3982	0.3980	0.3977	0.3973
0.1	0.3970	0.3965	0.3961	0.3956	0.3951	0.3945	0.3939	0.3932	0.3925	0.3918
0.2	0.3910	0.3902	0.3894	0.3885	0.3876	0.3867	0.3857	0.3847	0.3836	0.3825
0.3	0.3814	0.3802	0.3790	0.3778	0.3765	0.3752	0.3739	0.3725	0.3712	0.3697
0.4	0.3683	0.3668	0.3653	0.3637	0.3621	0.3605	0.3589	0.3572	0.3555	0.3538
0.5	0.3521	0.3503	0.3485	0.3467	0.3448	0.3429	0.3410	0.3391	0.3372	0.3352
0.6	0.3332	0.3312	0.3292	0.3271	0.3251	0.3230	0.3209	0.3187	0.3166	0.3144
0.7	0.3123	0.3101	0.3079	0.3056	0.3034	0.3011	0.2989	0.2966	0.2943	0.2920
0.8	0.2897	0.2874	0.2850	0.2827	0.2803	0.2780	0.2756	0.2732	0.2709	0.2685
0.9	0.2661	0.2637	0.2613	0.2589	0.2565	0.2541	0.2516	0.2492	0.2468	0.2444
1.0	0.2420	0.2396	0.2371	0.2347	0.2323	0.2299	0.2275	0.2251	0.2227	0.2203
1.1	0.2179	0.2155	0.2131	0.2107	0.2083	0.2059	0.2036	0.2012	0.1989	0.1965
1.2	0.1942	0.1919	0.1895	0.1872	0.1849	0.1826	0.1804	0.1781	0.1758	0.1736
1.3	0.1714	0.1691	0.1669	0.1647	0.1626	0.1604	0.1582	0.1561	0.1539	0.1518
1.4	0.1497	0.1476	0.1456	0.1435	0.1415	0.1394	0.1374	0.1354	0.1334	0.1315
1.5	0.1295	0.1276	0.1257	0.1238	0.1219	0.1200	0.1182	0.1163	0.1145	0.1127
1.6	0.1109	0.1092	0.1074	0.1057	0.1040	0.1023	0.1006	0.0989	0.0973	0.0957
1.7	0.0940	0.0925	0.0909	0.0893	0.0878	0.0863	0.0848	0.0833	0.0818	0.0804
1.8	0.0790	0.0775	0.0761	0.0748	0.0734	0.0721	0.0707	0.0694	0.0681	0.0669

(*Continued*)

Table B.4.   (*Continued*)

+/- z	0.00	0.01	0.02	0.03	0.04	0.05	0.06	0.07	0.08	0.09
1.9	0.0656	0.0644	0.0632	0.0620	0.0608	0.0596	0.0584	0.0573	0.0562	0.0551
2.0	0.0540	0.0529	0.0519	0.0508	0.0498	0.0488	0.0478	0.0468	0.0459	0.0449
2.1	0.0440	0.0431	0.0422	0.0413	0.0404	0.0396	0.0387	0.0379	0.0371	0.0363
2.2	0.0355	0.0347	0.0339	0.0332	0.0325	0.0317	0.0310	0.0303	0.0297	0.0290
2.3	0.0283	0.0277	0.0270	0.0264	0.0258	0.0252	0.0246	0.0241	0.0235	0.0229
2.4	0.0224	0.0219	0.0213	0.0208	0.0203	0.0198	0.0194	0.0189	0.0184	0.0180
2.5	0.0175	0.0171	0.0167	0.0163	0.0158	0.0154	0.0151	0.0147	0.0143	0.0139
2.6	0.0136	0.0132	0.0129	0.0126	0.0122	0.0119	0.0116	0.0113	0.0110	0.0107
2.7	0.0104	0.0101	0.0099	0.0096	0.0093	0.0091	0.0088	0.0086	0.0084	0.0081
2.8	0.0079	0.0077	0.0075	0.0073	0.0071	0.0069	0.0067	0.0065	0.0063	0.0061
2.9	0.0060	0.0058	0.0056	0.0055	0.0053	0.0051	0.0050	0.0048	0.0047	0.0046
3.0	0.0044	0.0043	0.0042	0.0040	0.0039	0.0038	0.0037	0.0036	0.0035	0.0034
3.1	0.0033	0.0032	0.0031	0.0030	0.0029	0.0028	0.0027	0.0026	0.0025	0.0025
3.2	0.0024	0.0023	0.0022	0.0022	0.0021	0.0020	0.0020	0.0019	0.0018	0.0018
3.3	0.0017	0.0017	0.0016	0.0016	0.0015	0.0015	0.0014	0.0014	0.0013	0.0013
3.4	0.0012	0.0012	0.0012	0.0011	0.0011	0.0010	0.0010	0.0010	0.0009	0.0009
3.5	0.0009	0.0008	0.0008	0.0008	0.0008	0.0007	0.0007	0.0007	0.0007	0.0006
3.6	0.0006	0.0006	0.0006	0.0005	0.0005	0.0005	0.0005	0.0005	0.0005	0.0004
3.7	0.0004	0.0004	0.0004	0.0004	0.0004	0.0004	0.0003	0.0003	0.0003	0.0003
3.8	0.0003	0.0003	0.0003	0.0003	0.0003	0.0002	0.0002	0.0002	0.0002	0.0002
3.9	0.0002	0.0002	0.0002	0.0002	0.0002	0.0002	0.0002	0.0002	0.0001	0.0001

Note: In Excel use: NORM.S.DIST(z, FALSE).

Table B.5. Single sampling plans for normal inspection (ISO 2859-1:1999).

Acceptance quality limit, AQL, in percent nonconforming items and nonconformities per 100 items (normal inspection)

Sample size code letter	Sample size	0.010	0.015	0.025	0.040	0.065	0.10	0.15	0.25	0.40	0.65	1.0	1.5	2.5	4.0	6.5	10	15	25	40	65	100	150	250	400	650	1000
		Ac Re	Ac Re	Ac Re	Ac Re	Ac Re	Ac Re	Ac Re	Ac Re	Ac Re	Ac Re	Ac Re	Ac Re	Ac Re	Ac Re	Ac Re	Ac Re	Ac Re	Ac Re	Ac Re	Ac Re	Ac Re	Ac Re	Ac Re	Ac Re	Ac Re	Ac Re
A	2	↓	↓	↓	↓	↓	↓	↓	↓	↓	↓	↓	↓	↓	↓	↓	↓	0 1	1 2	2 3	3 4	5 6	7 8	10 11	14 15	21 22	30 31
B	3	↓	↓	↓	↓	↓	↓	↓	↓	↓	↓	↓	↓	↓	↓	↓	0 1	1 2	2 3	3 4	5 6	7 8	10 11	14 15	21 22	30 31	44 45
C	5	↓	↓	↓	↓	↓	↓	↓	↓	↓	↓	↓	↓	↓	↓	0 1	1 2	2 3	3 4	5 6	7 8	10 11	14 15	21 22	30 31	44 45	↑
D	8	↓	↓	↓	↓	↓	↓	↓	↓	↓	↓	↓	↓	↓	0 1	1 2	2 3	3 4	5 6	7 8	10 11	14 15	21 22	30 31	44 45	↑	↑
E	13	↓	↓	↓	↓	↓	↓	↓	↓	↓	↓	↓	↓	0 1	1 2	2 3	3 4	5 6	7 8	10 11	14 15	21 22	30 31	44 45	↑	↑	↑
F	20	↓	↓	↓	↓	↓	↓	↓	↓	↓	↓	↓	0 1	1 2	2 3	3 4	5 6	7 8	10 11	14 15	21 22	30 31	44 45	↑	↑	↑	↑
G	32	↓	↓	↓	↓	↓	↓	↓	↓	↓	↓	0 1	1 2	2 3	3 4	5 6	7 8	10 11	14 15	21 22	30 31	44 45	↑	↑	↑	↑	↑
H	50	↓	↓	↓	↓	↓	↓	↓	↓	↓	0 1	1 2	2 3	3 4	5 6	7 8	10 11	14 15	21 22	30 31	44 45	↑	↑	↑	↑	↑	↑
J	80	↓	↓	↓	↓	↓	↓	↓	↓	0 1	1 2	2 3	3 4	5 6	7 8	10 11	14 15	21 22	30 31	44 45	↑	↑	↑	↑	↑	↑	↑
K	125	↓	↓	↓	↓	↓	↓	↓	0 1	1 2	2 3	3 4	5 6	7 8	10 11	14 15	21 22	30 31	44 45	↑	↑	↑	↑	↑	↑	↑	↑
L	200	↓	↓	↓	↓	↓	↓	0 1	1 2	2 3	3 4	5 6	7 8	10 11	14 15	21 22	30 31	44 45	↑	↑	↑	↑	↑	↑	↑	↑	↑
M	315	↓	↓	↓	↓	↓	0 1	1 2	2 3	3 4	5 6	7 8	10 11	14 15	21 22	30 31	44 45	↑	↑	↑	↑	↑	↑	↑	↑	↑	↑
N	500	↓	↓	↓	↓	0 1	1 2	2 3	3 4	5 6	7 8	10 11	14 15	21 22	30 31	44 45	↑	↑	↑	↑	↑	↑	↑	↑	↑	↑	↑
P	800	↓	↓	↓	0 1	1 2	2 3	3 4	5 6	7 8	10 11	14 15	21 22	30 31	44 45	↑	↑	↑	↑	↑	↑	↑	↑	↑	↑	↑	↑
Q	1 250	↓	↓	0 1	1 2	2 3	3 4	5 6	7 8	10 11	14 15	21 22	30 31	44 45	↑	↑	↑	↑	↑	↑	↑	↑	↑	↑	↑	↑	↑
R	2 000	↓	0 1	1 2	2 3	3 4	5 6	7 8	10 11	14 15	21 22	30 31	44 45	↑	↑	↑	↑	↑	↑	↑	↑	↑	↑	↑	↑	↑	↑

↓ = Use the first sampling plan below the arrow. If sample size equals, or exceeds, lot size, carry out 100 % inspection.

↑ = Use the first sampling plan above the arrow.

Ac = Acceptance number

Re = Rejection number

Table B.6.   Single sampling plans for tightened inspection (ISO 2859-1:1999).

Sample size code letter	Sample size	Acceptance quality limit, AQL, in percent nonconforming items and nonconformities per 100 items (tightened inspection)																																																		
		0.010		0.015		0.025		0.040		0.065		0.10		0.15		0.25		0.40		0.65		1.0		1.5		2.5		4.0		6.5		10		15		25		40		65		100		150		250		400		650		1 000	
		Ac	Re	Ac	Re	Ac	Re	Ac	Re	Ac	Re	Ac	Re	Ac	Re	Ac	Re	Ac	Re	Ac	Re	Ac	Re	Ac	Re	Ac	Re	Ac	Re	Ac	Re	Ac	Re	Ac	Re	Ac	Re	Ac	Re	Ac	Re	Ac	Re	Ac	Re	Ac	Re	Ac	Re	Ac	Re	Ac	Re
A	2	↓		↓		↓		↓		↓		↓		↓		↓		↓		↓		↓		↓		↓		↓		↓		↓		↓		0	1	1	2	2	3	3	4	5	6	8	9	12	13	18	19	27	28
B	3	↓		↓		↓		↓		↓		↓		↓		↓		↓		↓		↓		↓		↓		↓		↓		↓		0	1	1	2	2	3	3	4	5	6	8	9	12	13	18	19	27	28	41	42
C	5	↓		↓		↓		↓		↓		↓		↓		↓		↓		↓		↓		↓		↓		↓		↓		0	1	1	2	2	3	3	4	5	6	8	9	12	13	18	19	27	28	41	42	↑	
D	8	↓		↓		↓		↓		↓		↓		↓		↓		↓		↓		↓		↓		↓		↓		0	1	1	2	2	3	3	4	5	6	8	9	12	13	18	19	27	28	41	42	↑		↑	
E	13	↓		↓		↓		↓		↓		↓		↓		↓		↓		↓		↓		↓		↓		0	1	1	2	2	3	3	4	5	6	8	9	12	13	18	19	27	28	41	42	↑		↑		↑	
F	20	↓		↓		↓		↓		↓		↓		↓		↓		↓		↓		↓		↓		0	1	1	2	2	3	3	4	5	6	8	9	12	13	18	19	27	28	41	42	↑		↑		↑		↑	
G	32	↓		↓		↓		↓		↓		↓		↓		↓		↓		↓		↓		0	1	1	2	2	3	3	4	5	6	8	9	12	13	18	19	27	28	41	42	↑		↑		↑		↑		↑	
H	50	↓		↓		↓		↓		↓		↓		↓		↓		↓		↓		0	1	1	2	2	3	3	4	5	6	8	9	12	13	18	19	27	28	41	42	↑		↑		↑		↑		↑		↑	
J	80	↓		↓		↓		↓		↓		↓		↓		↓		↓		0	1	1	2	2	3	3	4	5	6	8	9	12	13	18	19	27	28	41	42	↑		↑		↑		↑		↑		↑		↑	
K	125	↓		↓		↓		↓		↓		↓		↓		↓		0	1	1	2	2	3	3	4	5	6	8	9	12	13	18	19	27	28	41	42	↑		↑		↑		↑		↑		↑		↑		↑	
L	200	↓		↓		↓		↓		↓		↓		↓		0	1	1	2	2	3	3	4	5	6	8	9	12	13	18	19	27	28	41	42	↑		↑		↑		↑		↑		↑		↑		↑		↑	
M	315	↓		↓		↓		↓		↓		↓		0	1	1	2	2	3	3	4	5	6	8	9	12	13	18	19	27	28	41	42	↑		↑		↑		↑		↑		↑		↑		↑		↑		↑	
N	500	↓		↓		↓		↓		↓		0	1	1	2	2	3	3	4	5	6	8	9	12	13	18	19	27	28	41	42	↑		↑		↑		↑		↑		↑		↑		↑		↑		↑		↑	
P	800	↓		↓		↓		↓		0	1	1	2	2	3	3	4	5	6	8	9	12	13	18	19	27	28	41	42	↑		↑		↑		↑		↑		↑		↑		↑		↑		↑		↑		↑	
Q	1 250	↓		↓		↓		0	1	1	2	2	3	3	4	5	6	8	9	12	13	18	19	27	28	41	42	↑		↑		↑		↑		↑		↑		↑		↑		↑		↑		↑		↑		↑	
R	2 000	↓		↓		0	1	1	2	2	3	3	4	5	6	8	9	12	13	18	19	27	28	41	42	↑		↑		↑		↑		↑		↑		↑		↑		↑		↑		↑		↑		↑		↑	
S	3 150	↓		0	1	1	2	2	3	3	4	5	6	8	9	12	13	18	19	27	28	41	42	↑		↑		↑		↑		↑		↑		↑		↑		↑		↑		↑		↑		↑		↑		↑	

⇩ = Use the first sampling plan below the arrow. If sample size equals, or exceeds, lot size, carry out 100 % inspection.

⇧ = Use the first sampling plan above the arrow.

Ac = Acceptance number

Re = Rejection number

Table B.7. Single sampling plans for reduced inspection (ISO 2859-1:1999).

Acceptance quality limit, AQL, in percent nonconforming items and nonconformities per 100 items (reduced inspection)

Each cell below shows the pair **Ac Re** (Acceptance number / Rejection number). ↓ = use the first sampling plan below the arrow; ↑ = use the first sampling plan above the arrow.

Sample size code letter	Sample size	0,010	0,015	0,025	0,040	0,065	0,10	0,15	0,25	0,40	0,65	1,0	1,5	2,5	4,0	6,5	10	15	25	40	65	100	150	250	400	650	1 000
A	2	↓	↓	↓	↓	↓	↓	↓	↓	↓	↓	↓	↓	↓	↓	0 1	1 2	↑	↑	↑	↑	↑	↑	↑	↑	↑	↑
B	2	↓	↓	↓	↓	↓	↓	↓	↓	↓	↓	↓	↓	↓	0 1	1 2	↑	↑	↑	↑	↑	↑	↑	↑	↑	↑	↑
C	2	↓	↓	↓	↓	↓	↓	↓	↓	↓	↓	↓	↓	0 1	1 2	↑	↑	↑	↑	↑	↑	↑	↑	↑	↑	↑	↑
D	3	↓	↓	↓	↓	↓	↓	↓	↓	↓	↓	↓	0 1	1 2	2 3	↑	↑	↑	↑	↑	↑	↑	↑	↑	↑	↑	↑
E	5	↓	↓	↓	↓	↓	↓	↓	↓	↓	↓	0 1	1 2	2 3	3 4	↑	↑	↑	↑	↑	↑	↑	↑	↑	↑	↑	↑
F	8	↓	↓	↓	↓	↓	↓	↓	↓	↓	0 1	1 2	2 3	3 4	5 6	7 8	↑	↑	↑	↑	↑	↑	↑	↑	↑	↑	↑
G	13	↓	↓	↓	↓	↓	↓	↓	↓	0 1	1 2	2 3	3 4	5 6	7 8	10 11	↑	↑	↑	↑	↑	↑	↑	↑	↑	↑	↑
H	20	↓	↓	↓	↓	↓	↓	↓	0 1	1 2	2 3	3 4	5 6	7 8	10 11	14 15	↑	↑	↑	↑	↑	↑	↑	↑	↑	↑	↑
J	32	↓	↓	↓	↓	↓	↓	0 1	1 2	2 3	3 4	5 6	7 8	10 11	14 15	21 22	30 31	↑	↑	↑	↑	↑	↑	↑	↑	↑	↑
K	50	↓	↓	↓	↓	↓	0 1	1 2	2 3	3 4	5 6	7 8	10 11	14 15	21 22	30 31	↑	↑	↑	↑	↑	↑	↑	↑	↑	↑	↑
L	80	↓	↓	↓	↓	0 1	1 2	2 3	3 4	5 6	7 8	10 11	14 15	21 22	30 31	↑	↑	↑	↑	↑	↑	↑	↑	↑	↑	↑	↑
M	125	↓	↓	↓	0 1	1 2	2 3	3 4	5 6	7 8	10 11	14 15	21 22	30 31	↑	↑	↑	↑	↑	↑	↑	↑	↑	↑	↑	↑	↑
N	200	↓	↓	0 1	1 2	2 3	3 4	5 6	7 8	10 11	14 15	21 22	30 31	↑	↑	↑	↑	↑	↑	↑	↑	↑	↑	↑	↑	↑	↑
P	315	↓	0 1	1 2	2 3	3 4	5 6	7 8	10 11	14 15	21 22	30 31	↑	↑	↑	↑	↑	↑	↑	↑	↑	↑	↑	↑	↑	↑	↑
Q	500	0 1	1 2	2 3	3 4	5 6	7 8	10 11	14 15	21 22	30 31	↑	↑	↑	↑	↑	↑	↑	↑	↑	↑	↑	↑	↑	↑	↑	↑
R	800	1 2	2 3	3 4	5 6	7 8	10 11	14 15	21 22	30 31	↑	↑	↑	↑	↑	↑	↑	↑	↑	↑	↑	↑	↑	↑	↑	↑	↑

↓ = Use the first sampling plan below the arrow. If sample size equals, or exceeds, lot size, carry out 100 % inspection.

↑ = Use the first sampling plan above the arrow.

Ac = Acceptance number

Re = Rejection number

Table B.8.   Double sampling plans for normal inspection (ISO 2859-1:1999).

Acceptance quality limit, AQL, in percent nonconforming items and nonconformities per 100 items (normal inspection)

Each cell shows **Ac Re** (Ac = Acceptance number, Re = Rejection number). ↓ = use first sampling plan below the arrow; ↑ = use first sampling plan above the arrow; * = use the corresponding single sampling plan.

Code	Sample	Sample size	Cum. sample size	0.010	0.015	0.025	0.040	0.065	0.10	0.15	0.25	0.40	0.65	1.0	1.5	2.5	4.0	6.5	10	15	25	40	65	100	150	250	400	650	1000
A	First			↓	↓	↓	↓	↓	↓	↓	↓	↓	↓	↓	↓	↓	↓	↓	↓	↓	↓	↓	↓	↓	↓	↓	↓	↓	↓
	Second			↓	↓	↓	↓	↓	↓	↓	↓	↓	↓	↓	↓	↓	↓	↓	↓	↓	↓	↓	↓	↓	↓	↓	↓	↓	↓
B	First	2	2	↓	↓	↓	↓	↓	↓	↓	↓	↓	↓	↓	↓	↓	↓	↓	↓	0 2	0 3	*	*	*	*	*	*	*	*
	Second	2	4	↓	↓	↓	↓	↓	↓	↓	↓	↓	↓	↓	↓	↓	↓	↓	↓	1 2	3 4	*	*	*	*	*	*	*	*
C	First	3	3	↓	↓	↓	↓	↓	↓	↓	↓	↓	↓	↓	↓	↓	↓	↓	0 2	0 3	1 4	*	*	*	*	*	*	*	↑
	Second	3	6	↓	↓	↓	↓	↓	↓	↓	↓	↓	↓	↓	↓	↓	↓	↓	1 2	3 4	4 5	*	*	*	*	*	*	*	↑
D	First	5	5	↓	↓	↓	↓	↓	↓	↓	↓	↓	↓	↓	↓	↓	↓	0 2	0 3	1 4	2 5	3 7	*	*	*	*	*	↑	↑
	Second	5	10	↓	↓	↓	↓	↓	↓	↓	↓	↓	↓	↓	↓	↓	↓	1 2	3 4	4 5	6 7	8 9	*	*	*	*	*	↑	↑
E	First	8	8	↓	↓	↓	↓	↓	↓	↓	↓	↓	↓	↓	↓	↓	0 2	0 3	1 4	2 5	3 7	5 9	*	*	*	*	↑	↑	↑
	Second	8	16	↓	↓	↓	↓	↓	↓	↓	↓	↓	↓	↓	↓	↓	1 2	3 4	4 5	6 7	8 9	12 13	*	*	*	*	↑	↑	↑
F	First	13	13	↓	↓	↓	↓	↓	↓	↓	↓	↓	↓	↓	↓	0 2	0 3	1 4	2 5	3 7	5 9	7 11	*	*	*	↑	↑	↑	↑
	Second	13	26	↓	↓	↓	↓	↓	↓	↓	↓	↓	↓	↓	↓	1 2	3 4	4 5	6 7	8 9	12 13	18 19	*	*	*	↑	↑	↑	↑
G	First	20	20	↓	↓	↓	↓	↓	↓	↓	↓	↓	↓	↓	0 2	0 3	1 4	2 5	3 7	5 9	7 11	11 16	17 22	*	↑	↑	↑	↑	↑
	Second	20	40	↓	↓	↓	↓	↓	↓	↓	↓	↓	↓	↓	1 2	3 4	4 5	6 7	8 9	12 13	18 19	26 27	37 38	*	↑	↑	↑	↑	↑
H	First	32	32	↓	↓	↓	↓	↓	↓	↓	↓	↓	↓	0 2	0 3	1 4	2 5	3 7	5 9	7 11	11 16	17 22	25 31	↑	↑	↑	↑	↑	↑
	Second	32	64	↓	↓	↓	↓	↓	↓	↓	↓	↓	↓	1 2	3 4	4 5	6 7	8 9	12 13	18 19	26 27	37 38	56 57	↑	↑	↑	↑	↑	↑
J	First	50	50	↓	↓	↓	↓	↓	↓	↓	↓	↓	0 2	0 3	1 4	2 5	3 7	5 9	7 11	11 16	17 22	25 31	↑	↑	↑	↑	↑	↑	↑
	Second	50	100	↓	↓	↓	↓	↓	↓	↓	↓	↓	1 2	3 4	4 5	6 7	8 9	12 13	18 19	26 27	37 38	56 57	↑	↑	↑	↑	↑	↑	↑
K	First	80	80	↓	↓	↓	↓	↓	↓	↓	↓	0 2	0 3	1 4	2 5	3 7	5 9	7 11	11 16	17 22	25 31	↑	↑	↑	↑	↑	↑	↑	↑
	Second	80	160	↓	↓	↓	↓	↓	↓	↓	↓	1 2	3 4	4 5	6 7	8 9	12 13	18 19	26 27	37 38	56 57	↑	↑	↑	↑	↑	↑	↑	↑
L	First	125	125	↓	↓	↓	↓	↓	↓	↓	0 2	0 3	1 4	2 5	3 7	5 9	7 11	11 16	17 22	25 31	↑	↑	↑	↑	↑	↑	↑	↑	↑
	Second	125	250	↓	↓	↓	↓	↓	↓	↓	1 2	3 4	4 5	6 7	8 9	12 13	18 19	26 27	37 38	56 57	↑	↑	↑	↑	↑	↑	↑	↑	↑
M	First	200	200	↓	↓	↓	↓	↓	↓	0 2	0 3	1 4	2 5	3 7	5 9	7 11	11 16	17 22	25 31	↑	↑	↑	↑	↑	↑	↑	↑	↑	↑
	Second	200	400	↓	↓	↓	↓	↓	↓	1 2	3 4	4 5	6 7	8 9	12 13	18 19	26 27	37 38	56 57	↑	↑	↑	↑	↑	↑	↑	↑	↑	↑
N	First	315	315	↓	↓	↓	↓	↓	0 2	0 3	1 4	2 5	3 7	5 9	7 11	11 16	17 22	25 31	↑	↑	↑	↑	↑	↑	↑	↑	↑	↑	↑
	Second	315	630	↓	↓	↓	↓	↓	1 2	3 4	4 5	6 7	8 9	12 13	18 19	26 27	37 38	56 57	↑	↑	↑	↑	↑	↑	↑	↑	↑	↑	↑
P	First	500	500	↓	↓	↓	↓	0 2	0 3	1 4	2 5	3 7	5 9	7 11	11 16	17 22	25 31	↑	↑	↑	↑	↑	↑	↑	↑	↑	↑	↑	↑
	Second	500	1000	↓	↓	↓	↓	1 2	3 4	4 5	6 7	8 9	12 13	18 19	26 27	37 38	56 57	↑	↑	↑	↑	↑	↑	↑	↑	↑	↑	↑	↑
Q	First	800	800	↓	↓	↓	0 2	0 3	1 4	2 5	3 7	5 9	7 11	11 16	17 22	25 31	↑	↑	↑	↑	↑	↑	↑	↑	↑	↑	↑	↑	↑
	Second	800	1600	↓	↓	↓	1 2	3 4	4 5	6 7	8 9	12 13	18 19	26 27	37 38	56 57	↑	↑	↑	↑	↑	↑	↑	↑	↑	↑	↑	↑	↑
R	First	1250	1250	↓	↓	0 2	0 3	1 4	2 5	3 7	5 9	7 11	11 16	17 22	25 31	↑	↑	↑	↑	↑	↑	↑	↑	↑	↑	↑	↑	↑	↑
	Second	1250	2500	↓	↓	1 2	3 4	4 5	6 7	8 9	12 13	18 19	26 27	37 38	56 57	↑	↑	↑	↑	↑	↑	↑	↑	↑	↑	↑	↑	↑	↑

⇩ = Use the first sampling plan below the arrow. If sample size equals, or exceeds, lot size, carry out 100 % inspection.

⇧ = Use the first sampling plan above the arrow.

Ac = Acceptance number

Re = Rejection number

* = Use the corresponding single sampling plan (or alternatively use the double sampling plan below, where available).

Table B.9. Cumulative Poisson probabilities.

np						x (Acceptance No.) c								
No. Def in s	0	1	2	3	4	5	6	7	8	9	10	11	12	13
0.05	0.951	0.999	1.000											
0.10	0.905	0.995	1.000											
0.15	0.861	0.990	0.999	1.000										
0.20	0.819	0.982	0.999	1.000										
0.25	0.779	0.974	0.998	1.000		P(X = <c)								
0.30	0.741	0.963	0.996	1.000										
0.35	0.705	0.951	0.994	1.000										
0.40	0.670	0.938	0.992	0.999	1.000									
0.45	0.638	0.925	0.989	0.999	1.000									
0.50	0.607	0.910	0.986	0.998	1.000									
0.55	0.577	0.894	0.982	0.998	1.000									
0.60	0.549	0.878	0.977	0.997	1.000									
0.65	0.522	0.861	0.972	0.996	0.999									
0.70	0.497	0.844	0.966	0.994	0.999	1.000								
0.75	0.472	0.827	0.959	0.993	0.999	1.000								
0.80	0.449	0.809	0.953	0.991	0.999	1.000								
0.85	0.427	0.791	0.945	0.989	0.998	1.000								
0.90	0.407	0.772	0.937	0.987	0.998	1.000								

(Continued)

Table B.9.   (*Continued*)

np

No. Def in s	x (Acceptance No.) c													
	0	1	2	3	4	5	6	7	8	9	10	11	12	13
0.95	0.387	0.754	0.929	0.984	0.997	1.000								
1.00	0.368	0.736	0.920	0.981	0.996	0.999	1.000							
1.10	0.333	0.699	0.900	0.974	0.995	0.999	1.000							
1.20	0.301	0.663	0.879	0.966	0.992	0.998	1.000							
1.30	0.273	0.627	0.857	0.957	0.989	0.998	1.000							
1.40	0.247	0.592	0.833	0.946	0.986	0.997	0.999	1.000						
1.50	0.223	0.558	0.809	0.934	0.981	0.996	0.999	1.000						
1.60	0.202	0.525	0.783	0.921	0.976	0.994	0.999	1.000						
1.70	0.183	0.493	0.757	0.907	0.970	0.992	0.998	1.000						
1.80	0.165	0.463	0.731	0.891	0.964	0.990	0.997	0.999	1.000					
1.90	0.150	0.434	0.704	0.875	0.956	0.987	0.997	0.999	1.000					
2.00	0.135	0.406	0.677	0.857	0.947	0.983	0.995	0.999	1.000					
2.20	0.111	0.355	0.623	0.819	0.928	0.975	0.993	0.998	1.000					
2.40	0.091	0.308	0.570	0.779	0.904	0.964	0.988	0.997	0.999	1.000				
2.60	0.074	0.267	0.518	0.736	0.877	0.951	0.983	0.995	0.999	1.000				
2.80	0.061	0.231	0.469	0.692	0.848	0.935	0.976	0.992	0.998	0.999	1.000			
3.00	0.050	0.199	0.423	0.647	0.815	0.916	0.966	0.988	0.996	0.999	1.000			
3.20	0.041	0.171	0.380	0.603	0.781	0.895	0.955	0.983	0.994	0.998	1.000			

Table B.9. (*Continued*)

np														
	x (Acceptance No.) c													
No. Def in s	0	1	2	3	4	5	6	7	8	9	10	11	12	13
3.40	0.033	0.147	0.340	0.558	0.744	0.871	0.942	0.977	0.992	0.997	0.999	1.000		
3.60	0.027	0.126	0.303	0.515	0.706	0.844	0.927	0.969	0.988	0.996	0.999	1.000		
3.80	0.022	0.107	0.269	0.473	0.668	0.816	0.909	0.960	0.984	0.994	0.998	0.999	1.000	
4.00	0.018	0.092	0.238	0.433	0.629	0.785	0.889	0.949	0.979	0.992	0.997	0.999	1.000	
4.20	0.015	0.078	0.210	0.395	0.590	0.753	0.867	0.936	0.972	0.989	0.996	0.999	1.000	
4.40	0.012	0.066	0.185	0.359	0.551	0.720	0.844	0.921	0.964	0.985	0.994	0.998	0.999	1.000
4.60	0.010	0.056	0.163	0.326	0.513	0.686	0.818	0.905	0.955	0.980	0.992	0.997	0.999	1.000
4.80	0.008	0.048	0.143	0.294	0.476	0.651	0.791	0.887	0.944	0.975	0.990	0.996	0.999	1.000
5.00	0.007	0.040	0.125	0.265	0.440	0.616	0.762	0.867	0.932	0.968	0.986	0.995	0.998	0.999
5.20	0.006	0.034	0.109	0.238	0.406	0.581	0.732	0.845	0.918	0.960	0.982	0.993	0.997	0.999
5.40	0.005	0.029	0.095	0.213	0.373	0.546	0.702	0.822	0.903	0.951	0.977	0.990	0.996	0.999
5.60	0.004	0.024	0.082	0.191	0.342	0.512	0.670	0.797	0.886	0.941	0.972	0.988	0.995	0.998
5.80	0.003	0.021	0.072	0.170	0.313	0.478	0.638	0.771	0.867	0.929	0.965	0.984	0.993	0.997
6.00	0.002	0.017	0.062	0.151	0.285	0.446	0.606	0.744	0.847	0.916	0.957	0.980	0.991	0.996
6.20	0.002	0.015	0.054	0.134	0.259	0.414	0.574	0.716	0.826	0.902	0.949	0.975	0.989	0.995
6.40	0.002	0.012	0.046	0.119	0.235	0.384	0.542	0.687	0.803	0.886	0.939	0.969	0.986	0.994
6.60	0.001	0.010	0.040	0.105	0.213	0.355	0.511	0.658	0.780	0.869	0.927	0.963	0.982	0.992
6.80	0.001	0.009	0.034	0.093	0.192	0.327	0.480	0.628	0.755	0.850	0.915	0.955	0.978	0.990

(*Continued*)

Table B.9. *(Continued)*

np No. Def in s	x (Acceptance No.) c													
	0	1	2	3	4	5	6	7	8	9	10	11	12	13
7.00	0.001	0.007	0.030	0.082	0.173	0.301	0.450	0.599	0.729	0.830	0.901	0.947	0.973	0.987
7.20	0.001	0.006	0.025	0.072	0.156	0.276	0.420	0.569	0.703	0.810	0.887	0.937	0.967	0.984
7.40	0.001	0.005	0.022	0.063	0.140	0.253	0.392	0.539	0.676	0.788	0.871	0.926	0.961	0.980
7.60	0.001	0.004	0.019	0.055	0.125	0.231	0.365	0.510	0.648	0.765	0.854	0.915	0.954	0.976
7.80	0.000	0.004	0.016	0.048	0.112	0.210	0.338	0.481	0.620	0.741	0.835	0.902	0.945	0.971
8.00	0.000	0.003	0.014	0.042	0.100	0.191	0.313	0.453	0.593	0.717	0.816	0.888	0.936	0.966
8.20	0.000	0.003	0.012	0.037	0.089	0.174	0.290	0.425	0.565	0.692	0.796	0.873	0.926	0.960
8.40	0.000	0.002	0.010	0.032	0.079	0.157	0.267	0.399	0.537	0.666	0.774	0.857	0.915	0.952
8.60	0.000	0.002	0.009	0.028	0.070	0.142	0.246	0.373	0.509	0.640	0.752	0.840	0.903	0.945
8.80	0.000	0.001	0.007	0.024	0.062	0.128	0.226	0.348	0.482	0.614	0.729	0.822	0.890	0.936
9.00	0.000	0.001	0.006	0.021	0.055	0.116	0.207	0.324	0.456	0.587	0.706	0.803	0.876	0.926
9.20	0.000	0.001	0.005	0.018	0.049	0.104	0.189	0.301	0.430	0.561	0.682	0.783	0.861	0.916
9.40	0.000	0.001	0.005	0.016	0.043	0.093	0.173	0.279	0.404	0.535	0.658	0.763	0.845	0.904
9.60	0.000	0.001	0.004	0.014	0.038	0.084	0.157	0.258	0.380	0.509	0.633	0.741	0.828	0.892
9.80	0.000	0.001	0.003	0.012	0.033	0.075	0.143	0.239	0.356	0.483	0.608	0.719	0.810	0.879
10.00	0.000	0.000	0.003	0.010	0.029	0.067	0.130	0.220	0.333	0.458	0.583	0.697	0.792	0.864
10.20	0.000	0.000	0.002	0.009	0.026	0.060	0.118	0.203	0.311	0.433	0.558	0.674	0.772	0.849
10.40	0.000	0.000	0.002	0.008	0.023	0.053	0.107	0.186	0.290	0.409	0.533	0.650	0.752	0.834

Table B.9. (*Continued*)

np No. Def in s	x (Acceptance No.) c													
	0	1	2	3	4	5	6	7	8	9	10	11	12	13
10.60	0.000	0.000	0.002	0.007	0.020	0.048	0.097	0.171	0.269	0.385	0.508	0.627	0.732	0.817
10.80	0.000	0.000	0.001	0.006	0.017	0.042	0.087	0.157	0.250	0.363	0.484	0.603	0.710	0.799
11.00	0.000	0.000	0.001	0.005	0.015	0.038	0.079	0.143	0.232	0.341	0.460	0.579	0.689	0.781
11.20	0.000	0.000	0.001	0.004	0.013	0.033	0.071	0.131	0.215	0.319	0.436	0.555	0.667	0.762
11.40	0.000	0.000	0.001	0.004	0.012	0.029	0.064	0.119	0.198	0.299	0.413	0.532	0.644	0.743
11.60	0.000	0.000	0.001	0.003	0.010	0.026	0.057	0.108	0.183	0.279	0.391	0.508	0.622	0.723
11.80	0.000	0.000	0.001	0.003	0.009	0.023	0.051	0.099	0.169	0.260	0.369	0.485	0.599	0.702
12.00	0.000	0.000	0.001	0.002	0.008	0.020	0.046	0.090	0.155	0.242	0.347	0.462	0.576	0.682
12.20	0.000	0.000	0.000	0.002	0.007	0.018	0.041	0.081	0.142	0.225	0.327	0.439	0.553	0.660
12.40	0.000	0.000	0.000	0.002	0.006	0.016	0.037	0.073	0.131	0.209	0.307	0.417	0.530	0.639
12.60	0.000	0.000	0.000	0.001	0.005	0.014	0.033	0.066	0.120	0.194	0.288	0.395	0.508	0.617
12.80	0.000	0.000	0.000	0.001	0.004	0.012	0.029	0.060	0.109	0.179	0.269	0.374	0.485	0.595
13.00	0.000	0.000	0.000	0.001	0.004	0.011	0.026	0.054	0.100	0.166	0.252	0.353	0.463	0.573
13.20	0.000	0.000	0.000	0.001	0.003	0.009	0.023	0.049	0.091	0.153	0.235	0.333	0.441	0.551
13.40	0.000	0.000	0.000	0.001	0.003	0.008	0.020	0.044	0.083	0.141	0.219	0.314	0.420	0.529
13.60	0.000	0.000	0.000	0.001	0.002	0.007	0.018	0.039	0.075	0.130	0.204	0.295	0.399	0.507
13.80	0.000	0.000	0.000	0.001	0.002	0.006	0.016	0.035	0.068	0.119	0.189	0.277	0.378	0.486
14.00	0.000	0.000	0.000	0.000	0.002	0.006	0.014	0.032	0.062	0.109	0.176	0.260	0.358	0.464

(*Continued*)

Table B.9. (*Continued*)

np No. Def in s	x (Acceptance No.) c													
	0	1	2	3	4	5	6	7	8	9	10	11	12	13
14.20	0.000	0.000	0.000	0.000	0.002	0.005	0.013	0.028	0.056	0.100	0.163	0.244	0.339	0.443
14.40	0.000	0.000	0.000	0.000	0.001	0.004	0.011	0.025	0.051	0.092	0.151	0.228	0.320	0.423
14.60	0.000	0.000	0.000	0.000	0.001	0.004	0.010	0.023	0.046	0.084	0.139	0.213	0.302	0.402
14.80	0.000	0.000	0.000	0.000	0.001	0.003	0.009	0.020	0.042	0.077	0.129	0.198	0.285	0.383
15.00	0.000	0.000	0.000	0.000	0.001	0.003	0.008	0.018	0.037	0.070	0.118	0.185	0.268	0.363
15.20	0.000	0.000	0.000	0.000	0.001	0.002	0.007	0.016	0.034	0.064	0.109	0.172	0.251	0.344
15.40	0.000	0.000	0.000	0.000	0.001	0.002	0.006	0.014	0.030	0.058	0.100	0.160	0.236	0.326
15.60	0.000	0.000	0.000	0.000	0.001	0.002	0.005	0.013	0.027	0.053	0.092	0.148	0.221	0.308
15.80	0.000	0.000	0.000	0.000	0.000	0.002	0.005	0.011	0.025	0.048	0.084	0.137	0.207	0.291
16.00	0.000	0.000	0.000	0.000	0.000	0.001	0.004	0.010	0.022	0.043	0.077	0.127	0.193	0.275
16.20	0.000	0.000	0.000	0.000	0.000	0.001	0.004	0.009	0.020	0.039	0.071	0.117	0.180	0.259
16.40	0.000	0.000	0.000	0.000	0.000	0.001	0.003	0.008	0.018	0.035	0.065	0.108	0.168	0.243
16.60	0.000	0.000	0.000	0.000	0.000	0.001	0.003	0.007	0.016	0.032	0.059	0.100	0.156	0.228
16.80	0.000	0.000	0.000	0.000	0.000	0.001	0.002	0.006	0.014	0.029	0.054	0.092	0.145	0.214
17.00	0.000	0.000	0.000	0.000	0.000	0.001	0.002	0.005	0.013	0.026	0.049	0.085	0.135	0.201
17.20	0.000	0.000	0.000	0.000	0.000	0.001	0.002	0.005	0.011	0.024	0.045	0.078	0.125	0.188
17.40	0.000	0.000	0.000	0.000	0.000	0.001	0.002	0.004	0.010	0.021	0.041	0.071	0.116	0.176
17.60	0.000	0.000	0.000	0.000	0.000	0.000	0.001	0.004	0.009	0.019	0.037	0.065	0.107	0.164

Table B.9. (Continued)

np

No. Def in s	0	1	2	3	4	5	6	7	8	9	10	11	12	13
					x (Acceptance No.) c									
17.80	0.000	0.000	0.000	0.000	0.000	0.000	0.001	0.003	0.008	0.017	0.033	0.060	0.099	0.153
18.00	0.000	0.000	0.000	0.000	0.000	0.000	0.001	0.003	0.007	0.015	0.030	0.055	0.092	0.143
18.20	0.000	0.000	0.000	0.000	0.000	0.000	0.001	0.003	0.006	0.014	0.027	0.050	0.085	0.133
18.40	0.000	0.000	0.000	0.000	0.000	0.000	0.001	0.002	0.006	0.012	0.025	0.046	0.078	0.123
18.60	0.000	0.000	0.000	0.000	0.000	0.000	0.001	0.002	0.005	0.011	0.022	0.042	0.072	0.115
18.80	0.000	0.000	0.000	0.000	0.000	0.000	0.001	0.002	0.004	0.010	0.020	0.038	0.066	0.106
19.00	0.000	0.000	0.000	0.000	0.000	0.000	0.001	0.002	0.004	0.009	0.018	0.035	0.061	0.098
19.20	0.000	0.000	0.000	0.000	0.000	0.000	0.000	0.001	0.003	0.008	0.017	0.032	0.056	0.091
19.40	0.000	0.000	0.000	0.000	0.000	0.000	0.000	0.001	0.003	0.007	0.015	0.029	0.051	0.084
19.60	0.000	0.000	0.000	0.000	0.000	0.000	0.000	0.001	0.003	0.006	0.013	0.026	0.047	0.078
19.80	0.000	0.000	0.000	0.000	0.000	0.000	0.000	0.001	0.002	0.006	0.012	0.024	0.043	0.072
20.00	0.000	0.000	0.000	0.000	0.000	0.000	0.000	0.001	0.002	0.005	0.011	0.021	0.039	0.066

np

No. Def in s	0	1	2	3	4	5	6	7	8	9	10	11	12	13
					x (Acceptance No.) c									

References

[1] ISO 9000:2015 Quality management systems. Fundamentals and vocabulary.

[2] Deming, W. (2000). *Out of the Crisis*, MIT Press.

[3] https://www.ukas.com/accreditation/about/how-to-get-ukas-accreditation/.

[4] Bucher, J. L. (2004). *Metrology Handbook*, ASQ Quality Press. ProQuest Ebook Central, https://ebookcentral.proquest.com/lib/portsmouth-ebooks/detail.action?docID=3002524.

[5] Morris, A. S. (2001). *Measurement and Instrumentation Principles*, Elsevier Science & Technology. ProQuest Ebook Central, https://ebookcentral.proquest.com/lib/portsmouth-ebooks/detail.action?docID=297105.

[6] Montgomery, D. (2009). *Introduction to Statistical Quality Control*, Sixth edition, USA: John Wiley & Sons.

[7] ISO 2859-1:1999 Sampling procedures for inspection by attributes — Part 1: Sampling schemes indexed by acceptance quality limit (AQL) for lot-by-lot inspection.

[8] BS 6001-1:1999 Sampling procedures for inspection by attributes.

Index